Robson Ruiz Olivoto

Mechanisms of action involved in the increase of basal tension in myocytes

Robson Ruiz Olivoto

Mechanisms of action involved in the increase of basal tension in myocytes

Cardiac myocyte contraction in the absence of calcium

ScienciaScripts

Imprint

Cover image: www.ingimage.com

This book is a translation from the original published under ISBN 978-613-9-67999-7.

Publisher:
Sciencia Scripts
is a trademark of
Dodo Books Indian Ocean Ltd. and OmniScriptum S.R.L publishing group

120 High Road, East Finchley, London, N2 9ED, United Kingdom
Str. Armeneasca 28/1, office 1, Chisinau MD-2012, Republic of Moldova, Europe
Printed at: see last page
ISBN: 978-620-8-19015-6

SUMMARY

I dedicate this research to a very important person in my life. She has been a companion in happy times and even more of a companion in difficult times, she has been fundamental in enabling me to complete this thesis, she has been and always will be my *"safe haven"*, my wife, Bruna Précoma.

ACKNOWLEDGMENTS

To my advisor, Prof. Dr. Rosalvo Tadeu Hochmuller Fogaça, for all his dedication, understanding and trust in me.

A special thanks to my master's advisor, Carlos Estevam Nolf Daminani, his dedication and teachings were of immeasurable importance during my training as a scientist, without them this would never have been possible, to you "CHIEF" my heartfelt thanks.

Profª . Drª . Ilana Kassouf Silva for her indispensable collaboration during this research.

To the undergraduate and postgraduate students with whom I worked.

To the professors of the Postgraduate Program in Cellular and Molecular Biology who, during their classes or in informal conversations, collaborated significantly with the development of my experiments.

To the staff of the animal house of the Federal University of Paranà, Biology sector.

SUMMARY

Eugenol is an essential oil extracted from cloves and is widely used in dentistry and the food industry. Literature data shows that in intact papillae, Eugenol at high concentrations (5 mM) induces an increase in resting tension. The same can be observed in preparations of permeabilized skeletal muscle cells, and these contractions are partially inhibited by 2-APB, an IP3R blocker. In this study, we evaluated whether Eugenol's ability to increase the resting tension of atrial muscle trabeculae is independent of Ca^2 +. Wistar rats (250-300g) were anesthetized with ethyl ether inhalation. After thoracotomy, the hearts were removed and perfused with Ringer's solution for proper dissection. After being dissected, the trabeculae were immersed in 2.5 ml vats (Ringer 28+1°C), carbonated (oxygen), pH 7.4. In the intact trabeculae technique, the atria were stimulated by rectangular pulses (0.5 Hz). In the *skinned* fibers technique, the trabeculae were dissected in Solution R (Table 1), then permeabilized with saponin and/or Triton X-100. In intact trabeculae, a concentration-response curve to Eugenol was performed in Normal Ringer's solution. The experiments with intact atrial trabeculae were repeated using blockers of Ca release channels^{2+} from the SR in the absence and presence of Eugenol, in normal Ringer's solution and zero-Ca Ringer's solution^{2+} In the presence of Procaine and Rianodine, blockers of Ca release^{2+} from the SR, both in normal Ringer's solution and zero-Ca Ringer's solution^{2+} , Eugenol was able to induce an increase in resting tension. Total inhibition of eugenol's ability to increase resting tension was possible with the use of BDM, an inhibitor of contractile protein cyclization. After depletion of the SR by caffeine in zero-Ca Ringer's solution^{2+} , eugenol was able to increase resting tension. The data suggest that the way in which eugenol acts is independent of Ca^{2+} stored in the SR, but is the result of an active state of cross-bridge cyclization. In permeabilized trabeculae, ruthenium red was unable to inhibit the increase in resting tension. In the same way as in intact trabeculae, BDM was able to reverse the resting tension induced by eugenol. Even in the presence of a high concentration of EGTA, eugenol was able to induce an increase in resting tension. When the trabeculae were permeabilized with Triton X-100, the resting tension was reversed, and the same was observed when the trabeculae were permeabilized with saponin for 80 minutes. Treatment with colchicine potentiated the effects of eugenol. The results of the arsenite experiments suggest that one of the ways in which eugenol acts is through the inhibition of p38 MAPk, and the ATP dosage experiments indicate that another possible way in which eugenol acts is through the reduction of this compound in solution. Taken together, these results indicate that eugenol does not act directly on contractile proteins; it acts via a Ca^2 + -independent pathway and seems to act via a cellular pathway, which may be associated with microtubules. As it seems, eugenol does not act through a single pathway; according to our results, it may act through the p38 MAPk pathway and via the reduction of free ATP.

Key words: eugenol, atrial trabeculae, strength in the absence of calcium.

INTRODUCTION

The ability to produce movement is one of the most surprising properties of living matter and is not exclusive to the muscle. There are a wide variety of cells capable of producing it. For example, the movements of leukocytes; the movements of hairs on the surface of some cells, etc. In all these cases, however, this production of movement is secondary and subordinate to another cellular function. However, in the muscle it is the main function, along with generating force. Thus, the function of muscle tissue is to produce movement and/or generate force (Cingolani et al., 2004).

According to their structure, contractile properties and function, three types of muscle can be distinguished: skeletal muscle, smooth muscle and cardiac muscle.

1 - Cell Structures Involved in the Excitation-Contraction Coupling Process (ECC) of the Cardiac Muscle

Most of the heart is made up of contractile muscle cells, also known as *myocytes* or, more specifically, *cardiomyocytes*. The rest is made up of pacemaker cells and conduction tissue (which are related to the generation and propagation of the heart's electrical activity), blood vessels and extracellular matrix (Brilla et al., 1991). Although there are different types of cells in the heart, it is the atrial and ventricular myocytes which, through their contraction, complete the filling of the heart with blood and propel it through the blood vessels, promoting circulation. Human ventricular myocytes are larger, measuring approximately 10 to 25 µm in diameter and 50 to 100 µm in length. Atrial myocytes, on the other hand, are smaller and are approximately 10 µm in diameter and around 20 µm in length (Gordon et al., 1966).

Cardiac myocytes vary in shape, from a simple ellipse to a cylinder with branches that are several times longer than they are wide. Adjacent myocytes are connected by structures called intercalated disks. A segment of these discs - the *gap junction* - forms a junction with low electrical resistance, which allows impulses to be transmitted quickly from cell to cell and implies a practically simultaneous contraction of all the fibers. This means that cardiac tissue, although not an anatomical syncytium, behaves like one from a functional point of view, and is therefore called a *functional syncytium* (Gordon et al., 1966).

Piasmatic Membrane (Sarcolemma) and Glycocalyx

In the early 1970s, electron microscope micrographs obtained by cryo-fracturing revealed the true three-dimensional arrangement of lipids and proteins within cell membranes. As a result of this observation, the fluid mosaic model of the membrane was proposed (Singer & Nicolson, 1972). The surface membrane of the cardiac myocyte, then characterized as a lipid bilayer, is lipoprotein in nature.

For physiologists, classifying membrane proteins by their function is much more useful than by their structure. Cell membrane proteins function as: (1) *structural proteins*, (2) *enzymes*, (3) *receptors* and (4) *transporters*, which move molecules in and out of the cell.

The heart cell also has an outer layer called the glycocalyx. This layer is mainly made up of complex carbohydrates called polysaccharides, often associated with proteins (glycoproteins). The latter have a negative electrical charge with the property of binding to positively charged ions such as calcium, sodium and magnesium ions (Bennett, 1963). Therefore, the lipid bilayer is immediately next to the glycocalyx and is usually related as the only real component of the membrane (sarcolemma), thus constituting a bilayer contour of phospholipid molecules (Curry, 1986).

1.2- Tùbulos-T

In the sarcolemma of cardiac myocytes, we find invaginations that penetrate the cells, running transversely and giving off branches. These invaginations are called *Transverse Tubules* or *"T" Tubules.* On the other hand, other small depressions in the membrane cannot be characterized as such and are called *Caverns*. Because the T-tubules are an extension and have the same ultrastructure as the sarcolemma, they increase the cell's surface area by around 30%, making it easier for the excitatory stimulus to spread into the cell (Scales, 1981). The ultrastructural organization of the cardiac sarcolemma is important because this is the site through which calcium enters and leaves the cell. Thus, the location of the relevant transport systems is of fundamental importance. This is particularly the case if there is a differentiated distribution of ion channels, pumps and other membrane specializations (Almers & Stirling, 1984).

1.3- Sarcoplasmic Reticulum (SR)

Anatomically, the *sarcoplasmic reticulum (SR)* is a network spread across cardiac myocytes and demarcated by a lipid bilayer similar to that of the sarcolemma (Page et al., 1971). The main function of this organelle seems to be the sequestration and release of calcium into the cytoplasm (Winegrad, 1965). The ends of the RS are located close to and in apposition to the T-tubules. There, the RS tubules expand into pocket-like projections, which extend along the inner surface of the sarcolemma or are folded around the T-tubules. These expanded areas of the SR are called *subsarcolemmal cisterns* or *junctional components*. The connection between the SR and the transverse tubule is called the *triad*, while the connection between the SR and the cavern is called the *diad* (Franzini-Armstrong & Protasi,

1997). Except for the junctions between the SR and the sarcolemma, the sarcoplasmic reticulum membrane appears homogeneous and contains mainly the calcium-ATPase pump (SERCA) (Stewart & MacLenann, 1974; Katz et al., 1986). Thus, most of the surface of the sarcoplasmic reticulum seems to function basically for the removal of calcium from the cytoplasm (Franzini-Armstrong, 1975).

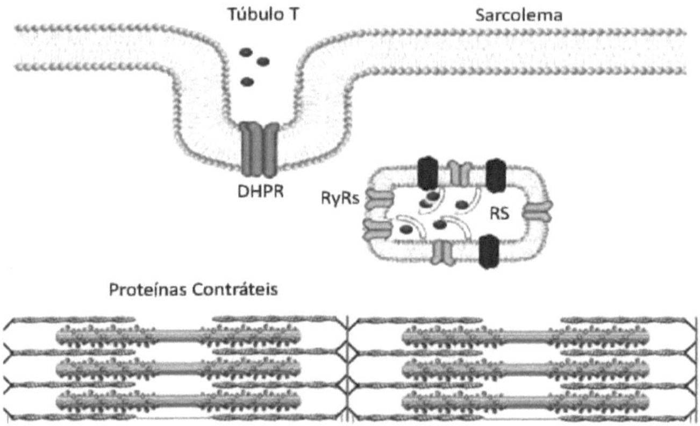

Figure 1 - Schematic structure of the T-tubule and the sarcoplasmic reticulum. The reticulum

The sarcoplasmic reticulum comprises each myofibril. The T-tubule system is closely associated with the sarcoplasmic reticulum (Source: Olivoto).

1.4- Sarcomere and Myofilaments

The main function of the cardiac myocyte is the process of muscle contraction and relaxation. The *sarcomere* is considered to be the basic contractile unit of the myocyte and is bounded by two *Z lines* or *disks*.

Highly specialized structures called *Myofilaments* or *Contractile Filaments* contribute to the performance of these functions. In turn, these myofilaments are made up of thick filaments containing *myosin* and thin filaments containing *actin*, as well as the contractile components of the cytoskeleton (Page *et al.,* 1971; Page, 1978).

Each thick filament is made up of -300 myosin molecules, but also contains other proteins such as *titin* and *protein-C*. Each myosin heavy chain (MW -450,000) has a long (~130nm) a-helical tail and a globular head (Warshaw, 1996). The tails of the myosin heavy chain form the main axis of the thick filament. These myosin heads form the *cross-bridges* that interact with actin to generate contraction (Offer, 1972; Freiburg & Gautel, 1996). These heads contain sites for ATP hydrolysis, and have two light chains associated with each head. Based on their susceptibility to specific proteases, the myosin heavy chain has been identified as *light meromyosin* (*LMM*, 2/3 of the 150 nm tail) and *heavy meromyosin (HMM). The* central region of the thick filament is devoid of cross-bridges, reflecting the tail-to-tail relationship of the myosin molecules (Warshaw, 1996). The two myosin light chains (*MLC1* and *MLC2*) bind to each heavy chain at the base of the S1 domain or neck region. MLC1 is also referred to as the *alkali* or *essential light chain* (*ELC*). MLC2 is also called the *phosphorylatable* or *regulatory light chain* (*RLC*). Both light chains confer physical stability to the thick filament. However, the regulatory light chain (MLC2) can also alter its function in response to Ca binding^{2+} or phosphorylation (Warshaw, 1996). The thin filament proteins and their interactions have been reviewed by Zot & Potter (1987); and by Solaro & Rarick (1998).

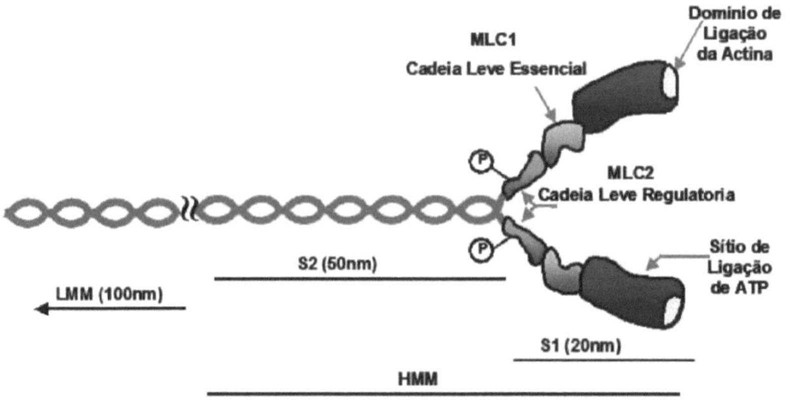

Figure 2 - Myofilaments: Myosin. The myosin molecule is approximately 170 nm long and has two globular heads (S1) and tails (including the light meromyosin LMM + S2), which are presented as an intertwined tail. The two light chains (MLC) are indicated in the neck region (Source: Warshaw, 1996).

The thin filament is made up of two chains of the globular *G-actin* protein which forms the double helix of the *F-actin* polymer (Solaro & Rarick, 1998; Rayment *et al.,* 1993 a, b; Bers, 2001). *Tropomyosin* (*Tm*) is a long, flexible protein that lodges in the cleft formed by the actin filaments and involves around seven actin monomers. Tropomyosin is also a double filament forming an α-helix and its two filaments may be connected by disulfide bridges (Lehrer *et al.,* 1997). Every seventh actin has a *Troponin* (*Tn*) complex attached to tropomyosin. This troponin complex is made up of 3 subunits: **a)** *Troponin T* (*TnT*, or tropomyosin-binding subunit), **b)** *Troponin C (TnC,* or Ca-binding subunit^{2+}) and **c)** *Troponin I* (*TnI*, or inhibitory subunit, which also binds to actin). TnT has a globular carboxyl region and an elongated shape that extends along Tm over about 3 actin monomers (Ohtsuki, 1979, Flicker *et al.,* 1982). TnI (near its amino terminus) interacts specifically with the carboxyl termini of both TnT and TnC. The strong binding of the amino termini of TnI and TnC depends on the binding of calcium or magnesium to the sites on the carboxyl terminus of TnC. In the resting condition (low $[Ca^{2+}]_i$) the terminal carboxyl of TnI also binds specifically to actin, and this prevents the myosin head from interacting with actin. When Ca^{2+} binds to the amino terminus of TnC, in the regulatory site with the lowest physiological affinity, this part of TnC binds to the carboxy terminus of TnI, inducing the dissociation of TnI from actin. As a result, TnT induces a movement of tropomyosin which, in turn, allows myosin to interact with actin (Solaro & Rarick, 1998; Rayment *et al.,* 1993 a, b).

In electron microscopy, the place where the actin and myosin filaments overlap is called *band A* (*anisotropic*), which appears dark in microscopy; in contrast to the lighter zones or *bands I* (*isotropic*) on either side, which contain only thin filaments. In the center of band A is an area known as *zone H*. Here only myosin is present, because the overlap between the thin and thick filaments ends near this zone. Each H zone contains a dark central region, the *M-line.* The thick filaments are interconnected transversely at the M-line by the *-M protein* and *myomesin* (Rayment et al., 1993a; Obermann et al., 1995). Another protein called *titin* has an extremely long structure and protrudes through the thick filament from the M-line to the Z-line (Gautel & Goulding, 1996).

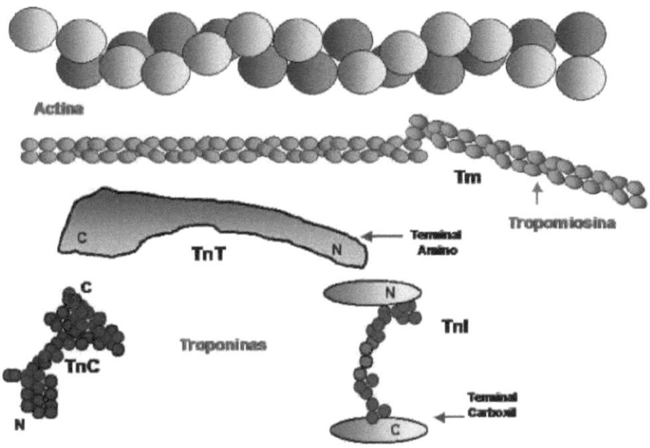

Figure 3 - Myofilament: Actin. Representative drawing of the actin structural complex (Source: Solaro & Rarick, 1998).

2 - Regulation of Contractile Protein Activity

The primary regulation of contraction in mammalian striated and cardiac muscle occurs in the thin filament. When free cytosolic calcium rises, it binds to the troponin complex, which in turn alters the thin filament to allow myosin to combine with actin and generate tension.

The occupation of troponin C by calcium strengthens the bond between troponin C and troponin I and thus reduces the negative interaction of troponin I with actin, causing a molecular change in tropomyosin, which becomes less inhibitory to actin-myosin interaction (Brenner, 1988; Rüegg, 1990).

According to Brenner (1988), it is predominantly the frequency of cross-bridge bonds that is affected by calcium. Thus, any increase in the concentration of free calcium in the myoplasm could increase the occupancy of troponin C by calcium which, in turn, increases the probability of cross-bridge bonds. Therefore, the first effect of calcium ions could be an increase in the force developed (Rüegg, 1990).

In contrast, Hancock et al. (1993) proposed that calcium ions interacting with troponin C act more by increasing the availability of binding sites for myosin on the actin filament than by altering the frequency of transition between cross-bridge states. In fact, these two apparently divergent proposals would not be so different, since calcium promotes the binding of myosin to actin, an effect which also alters the molecular configuration of the myosin head and, consequently, the degree of myosin-actin interaction (Rayment et al., 1993).

Normally, the myoplasmic concentration of calcium ions during systole is such that 50% of the contractile sites are activated. Thus, the heart still has a considerable contractile reserve, which could be exploited by increasing the occupation of troponin C by calcium. The development of force could ultimately depend on the amount of calcium released from the sarcoplasmic reticulum and offered to troponin C, which could explain the increase in contraction force during catecholamine stimulation (Opie, 1995). β-adrenergic stimulation acts either indirectly through cAMP or more directly by increasing the cytosolic concentration of calcium through an increase in the Ca^{2+} current and the release of Ca^{2+} from the SR. An increase in the frequency of cross-bridge bonds increases the speed of tension development, dF/dt. This is the positive inotropic effect. The speed of relaxation, (-dF/dt) will also be increased. This lusitropic effect is explained by two factors: firstly, phosphorylation of phospholamban activates calcium transport into the sarcoplasmic reticulum in such a way that free cytosolic calcium is decreased more rapidly in the presence of catecholamines (Katz, 1979) and, secondly, β-adrenergic activation decreases the sensitivity of the myofilament to calcium due to the phosphorylation of TnI and the consequent reduction in the affinity of TnC for calcium (Katz, 1979; Li *et al.,* 2000).

The dissociation rate constant of the troponin C- Ca complex^{2+} is greatly increased when troponin I is phosphorylated by PKA (Holroyde *et al.*, 1980). The physiological significance of this phosphorylation is the fact that contraction can be deactivated more quickly when calcium decreases at the beginning of diastole, which contributes to increased relaxation. Troponin T can also develop a similar phosphorylation in response to stimulation by protein kinase C (PKC), the latter being formed in the signaling system triggered by α_1-adrenergic stimulation. This phosphorylation of troponin T provides inhibitory results similar to those obtained by the phosphorylation of troponin I (Holroyde *et al.,* 1980).

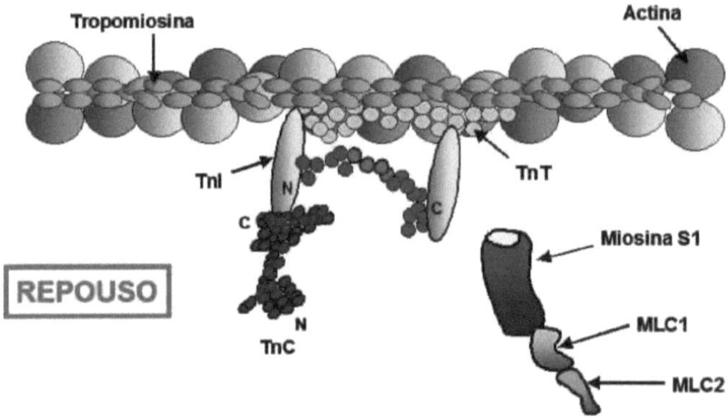

Figure 4 - The resting condition. In the resting condition, the carboxy-terminal of TnI is bound to actin, thus anchoring the TnT-tropomyosin complex (Tm) and preventing the myosin head (S1) from binding to actin (Source: Solaro & Rarick, 1998).

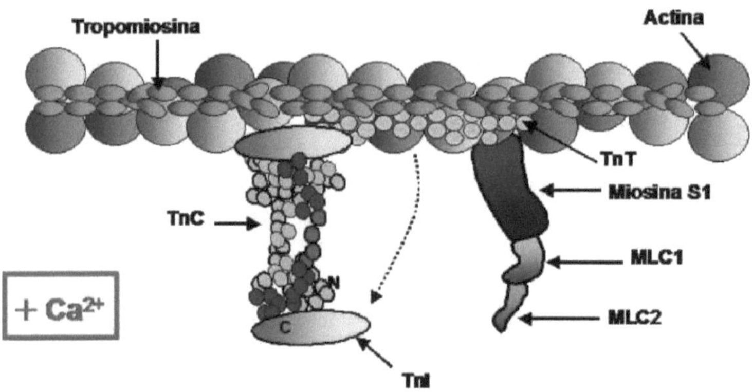

Figure 5 - Myofilaments in the Presence of Ca^{2+} . When Ca^{2+} binds to the amino terminal of TnC, this region binds strongly to the carboxyl terminal of TnI, which moves away from the actin, allowing the TnT-Tm complex to move to the bottom of the gap formed by the actin and, consequently, expose the sites located along the actin which will interact with the binding sites of the subfragment of the S1 heads of myosin. (Source: Solaro & Rarick, 1998).

3 - Molecular Biology of Cardiac Myocyte Calcium Channels

Calcium ions are important intracellular messengers in the cardiovascular system. Calcium homeostasis is regulated through: (1) an extracellular cycle, which controls the entry and removal of calcium between the cytosol and the extracellular space; (2) and an intracellular cycle, which controls flows between the cytosol and the sarcoplasmic reticulum. Calcium signaling in the cardiovascular system is mediated by several families of proteins. These processes include proteins that regulate the entry of calcium into the cytosol, proteins that recognize calcium as it enters the cytosol and proteins that remove calcium from the cytosol. The proteins that regulate the flow of calcium into the cytosol include the intracellular calcium release channels and the plasma membrane calcium channels (Katz, 1996a, b).

3.1- Proteins that Regulate Cellular Calcium Influx

Two families of proteins regulate the entry of calcium into the cytosol: **a)** one family includes the intracellular calcium release channels (from the RS) and, **b)** the other family constitutes the plasma membrane calcium channels.

1) Plasma membrane calcium channels

Calcium enters the cell through voltage-dependent calcium channels located in the plasma membrane, which participate in the extracellular calcium cycle. The L-type calcium channels in the plasma membrane are the most abundant members of this family in cardiac myocytes and contribute to both *plateau* formation and excitation-contraction coupling (Katz, 1996a, b) (Tsien & Tsien, 1990).

The calcium channels of the plasma membrane are members of a large family of ion channels which includes sodium channels, potassium channels and chloride channels. All these channels share a very similar structure.

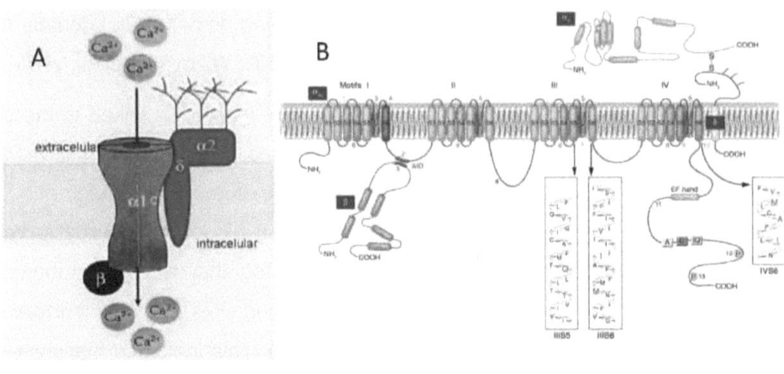

Figure 6 - (A) The calcium channel. Representative drawing of a slow calcium channel showing its respective subunits surrounding the central pore (Source: Perez - Reyez & Schneider, 1995). (B) Structure of Ca slow channels^{2+} (CACNA1C). Subunits a1, a2, β and δ. Subunit al is made up of four domains, each composed of six transmembrane segments. Subunit α2 is non-cytosolic and linked to subunit δ by disulfide bridges. δ subunit with a single transmembrane segment. Subunit with cytosolic arrangement, linked by a cytosolic peptide interaction domain to the al subunit between *domains* I and II (Source: Bodi et al., 2005).

a) Dihydropyridine Receptors (DHPRs)

In the 1970s, with the use of electron microscopy methods, it was possible to identify a transmembrane protein classified as a heterotetrameric polypeptide (Block et al., 1988) composed of five subunits: a1, a2, δ, β and, in some tissues, the γ subunit (Bodi et al., 2005). In skeletal muscle, the slow Ca channels^{2+} (CACNA1C) are formed by the αis, a2, δ, β and γ subunits. However, in cardiac muscle the channel is formed only by the α1c, a2, δ and β subunits, with the al subunit being added to the letter "c" to define cardiac muscle channels and the letter *"s"* to define skeletal muscle channels, and added to the letter (Bodi et al., 2005).

The a1c subunit (170-240 kDa) consists of 4 (four) homologous *domains* (I-IV), with each *domain* made up of 6 (six) transmembrane segments in an a-helix, defined as S1, S2, S3, S4, S5 and S6. Between the S3 and S4 segments is the peptide sequence responsible for the kinetic activation of the channel. This peptide sequence has a voltage-dependent characteristic and the S4 segment is considered to be the voltage sensor.

The S5 and S6 segments, when activated by the change in membrane voltage, undergo a change in their protein conformation with the consequent formation of an ion-selective channel, more specifically, for Ca ions^{2+} . The S6 segment contains the peptide sequence responsible for inactivating the channel (Tsien & Tsien, 1990; Cheng et al., 1993). Communication between the *domains* occurs through a cytosolic peptide bond between S6 and S1 of the next domain. The linker peptide sequence between domain I and domain II also acts as a linker between the aio and β subunits (Bodi et al., 2005; Yokoyama et al., 2005).

The a2 subunit, which is located extracellularly (Ahlijamian et al., 1990), is linked to the δ subunit by interaction on the cell surface through disulfide bridges. The δ subunit is formed by a single transmembrane segment next to *domain* IV of the aic subunit (De Jongh et al., 1989). Incorrect expression of these subunits can lead to the development of bradycardia (Ivanov et al., 2004). When the α2 and δ subunits are expressed together with the aic subunit, there is an increase in Dihydropyridine Receptor (DHPR) binding sites and ionic currents (Singer et al., 1991; Wei et al., 1995; Bangalore et al., 1996). This indicates that these subunits play an important role in the formation and function of these channels on the membrane surface (Bers, 2001) by increasing the affinity of the channel to the dihydropyridine compound (Wei et al., 1995).

Four isoforms of the subunit have been described (β1 - β4), however the isoform found in cardiac muscle is β2 (De Waard et al., 1994; Perez-Reyes & Schneider, 1995). They are hydrophilic, non-glycosylated structures located in the cytosol; they have no transmembrane segments and bind to the α_1 subunit between the I and II *domains* (De Warrd et al., 1994).

11- Sarcoplasmic reticulum membrane calcium channels

a) Rianodine Receptors (RyRs)

Anatomically, the sarcoplasmic reticulum (SR) is a network spread throughout the myocytes and demarcated by a lipid bilayer similar to that of the sarcolemma (Page et al., 1971). The main function of this organelle seems to be the sequestration and release of calcium into the cytoplasm (Winegrad, 1965).

In the 1980s, using electron microscopy techniques, it was possible to identify the presence of transmembrane protein structures arranged in the RS (Dulhunty, 2006). Using an alkaloid naturally found in the stem of the plant *Ryanna speciosa*, called ryanodine, it was possible to identify the function of these structures (Elison & Jenden, 1967; Jenden & Fairhurst, 1969). Experiments using this alkaloid compound induced a powerful paralysis of muscle contractile activity, but with different effects on skeletal and cardiac muscles. The action of this compound on skeletal muscle induced a total paralysis of contraction capacity, but this did not occur in cardiac muscle (Elison & Jenden, 1967; Fairhurst, 1973).

Based on their morphological differences and high affinity for the alkaloid ryanodine, these proteins, once purified, were then identified in muscle as Ca^2 + releasing channels of the RS (ryanodine receptors, RyR) (Lai et al., 1988; Inui et al., 1987a, b). Three different isoforms were initially identified, RyR1 in skeletal muscle, RyR2 in cardiac muscle, RyR3 which was initially identified in brain cells, but was later also found in other tissues (Meissner, 1986; Inui et al., 1987).

The structures identified by Franzini-Armstrong as feet are then identified as the DHPR and RyR. Block et al. (1988) demonstrated that these feet are organized in a distinct pattern in the SR immediate to the T-tubule membrane and that they are in apposition to an organized arrangement of particles in the T-tubule membrane similar to the Ca channel protein^{2+} of the sarcolemma (dihydropyridine receptor, *DHPR*). This arrangement in mammalian cardiac muscle has a RyR:DHPR stoichiometry of 4:10 depending on the species (Bers & Stiffel, 1993; Wibo et al., 1991). It is also important to note that in skeletal muscle, dihydropyridine receptors do not function as Ca channels^{2+} but act, during depolarization, as an element that promotes an increase in the permeability of the ryanodine channels of the SR. Furthermore, in skeletal muscle, the volume of the SR is significantly greater than in the myocardium, which explains the great independence of contractile function in relation to extracellular Ca^2 + (Bers & Stiffel, 1993; Wibo et al., 1991).

b) IP3 receptors (IP3Rs)

The membrane of the RS is not only home to RyR expression. IP3Rs (Inositol 1,4,5-triphosphate Receptors or InsP3Rs) are also expressed in its membrane. Inositol triphosphate receptors are predominant in smooth muscle where they mediate drug-mechanical coupling. In this process, the drug that binds to the membrane receptors releases the activating calcium that initiates contraction.

The IP3 receptor is very similar to the ryanodine receptor (RyR) and together they form a superfamily of intracellular Ca-releasing channels^{2+} . The IP3R was initially isolated from neural tissue (Supattapone et al., 1988) and smooth muscle (Chadwick et al., 1990). Cloning of the neuronal IP3R showed the existence of a sequence homologous to RyRs (Furuichi et al., 1989; Mignery et al., 1989). In 1990, Chadwick et al. showed that IP3Rs have a quaternary structure like skeletal muscle RyRs (despite their lower molecular weight) and, in addition, it was suggested that they have a tetrameric arrangement. There are 3 subtypes of IP3R: type 1; type 2 and type 3. Type 1 IP3R is predominantly present in smooth muscle and neuronal tissue. Types 2 and 3 have also been cloned and are homologous to type 1 by 69 and 64% respectively (Südhof et al., 1991; Blondel et al., 1993). Studies have examined the expression and subcellular localization of IP3R isoforms, as well as investigating their potential role in modulating the cardiac muscle AEC process. Western blot, PCR and IP3-ligation analyses indicate that both atrial and ventricular myocytes express mainly type 2 IP3Rs, with a population of approximately 6 times more IP3Rs in atrial cells. In the same study, the authors demonstrated that these channels are functional and located in the same cellular regions as the junctional RyRs, which are involved in the AEC process (Lipp et al., 2000).

In 1984, Hirata et al. were the first to demonstrate the release of Ca^2 + induced by Ca^2 + in cardiac muscle SR vesicles. However, other authors have shown no effect of IP3 either on isolated cardiac muscle SR or on myocytes permeabilized by saponin. In 1986, Nosek et al. demonstrated that IP3 was capable of potentiating both the spontaneous release of Ca^2 + and that induced by caffeine in cardiac muscle. Fabiato (1986a, b; 1990) showed that the rapid application of IP3 to dismembered rat myocytes could induce the release of Ca^{2+} , but that this was much smaller and slower than the release of Ca^{2+} by the CICR process. These data were confirmed in studies with dismembered rat ventricular muscle fibers (Kentish et al., 1990).

The biological role of this family of receptors in cardiac tissues is poorly understood. Although the IP3 receptor is not the main Ca^2 + releasing channel in cardiac muscle, IP3 may be involved in regulating Ca^{2+} release induced by Ca^{2+} . The increase in IP3 and the resulting increase in the cytosolic concentration of Ca^{2+} may be an important mechanism in the control of cardiac contractile force in response to hormonal and pharmacological factors as well as in pathological states (Go et al., 1995).

In the heart muscle, the ryanodine receptor acts as the main Ca release channel^{2+} of the SR in the excitation-contraction coupling process (Smith et al., 1989; Gyorke & Fill, 1993). It has been shown that the activation of phosphatidylinositol 4, 5-biphosphate, which is coupled to a- adrenergic and muscarinic receptors in the plasma membrane, is capable of increasing the level of IP3 in cardiac muscle (Scholz et al., 1988). Initial studies revealed that in a vesicular preparation of RS, IP3 was capable of inducing a slow release of Ca^2 + as well as activating contraction in preparations with dismembered rat ventricular muscle fiber and chicken heart (Hirata et al., 1984).

IP3Rs have three transmembrane domains, a binding site, a coupling site and a channel-forming site (Mignery & Südhof, 1990; Südhof et al., 1991). The sequence and domains are similar in the three types, but they are regulated differently by InsP3 and Ca^{2+} . Type 2 IP3Rs, which are more expressed in atrial myocytes, have a high affinity for the InsP3 compound (Perez et al., 1997; Ramos-Franco et al., 1998), while type 1 IP3Rs have 4 times less affinity and type 3 IP3Rs have approximately 10 times less affinity (Hagar & Ehrlich, 2000). IP3R activation is also sensitive to Ca^2 + however, activation mediated by Ca^2 + is differentially associated with type 1 isoforms, while type 2 and 3 IP3Rs are not activated by Ca^{2+} (Bezprozvanny et al., 1991). Type 2 IP3Rs, which are highly expressed in cardiac myocytes, especially atria, are much more sensitive to InsP3 and almost insensitive to physiological changes in the concentration of Ca^2 + ions, thus acting more like an InsP3 sensor in cardiac myocytes (Bers, 2001).

The conductance capacity of the IP3R channel is approximately half that found in RyR channels, but its selectivity for bivalent cations, such as Ca ions^2 +, is similar to that found in RyR channels (Bezprozvanny & Ehrlich, 1995). A powerful inhibitory regulator of the activation of these channels is heparin, which is able to alter the protein conformation of the binding sites, preventing sensitization via InsP3. On the other hand, cAMP-dependent proteins (PKA), which are phosphorylative, are capable of increasing the sensitization of IP3R channels induced by InsP3, although more is not known about other phosphorylation proteins (Burgess et al., 1991). In addition, FKBP is able to bind to InsP3 binding sites, decreasing the sensitization of IP3R channels, thus forming an FKBP-IP3R calcineurin complex (Cameron et al., 1995). Calmodulin is another protein that regulates the activity of IP3R channels, with binding sites in types 1 and 2 (Yamada et al., 1995).

The binding of InsP3 to IP3R channels occurs in the transmembrane domain called the *binding site*, which was initially characterized from its amino terminal (NH2) by Mignery et al. (1989) and based on the discovery of 410 residues with complementary characteristics to InsP3 (Moraru et al., 1999). Experiments carried out after Mignery et al.'s discovery indicated that this amino terminal feature was particular to all IP3R isoforms (Sudhof et al., 1991; Maranto, 1994; Morris et al., 2002). In 1996, Yoshikawa et al. identified a sequence of 10 arginine and lysine residues, distributed throughout the domain, as amino acids with a very important role in the binding of InsP3 to the IP3R channel, especially the Arg-265, Lys-508 and 511 residues, perhaps the latter being the most important for this binding.

The *pore* of IP3R channels was established by immunocytochemical techniques and by full-length N-glycosylation analysis (Galvan et al., 1999; Michikawa et al., 1994). The data from these studies, together with an analogy between the IP3R channel model and RyR channels, suggested that the carboxy-terminal (COOH) α-helix was involved in ion permeability. The interaction between InsP3 and its binding site in the IP3R channel induces changes in the protein conformation of the transmembrane segments in α-helices 5 and 6. This alteration is capable of creating the basic structure for pore formation (Michikawa et al., 1994; Williams et al., 2001; Taylor, Fonseca and Morris, 2004). In 1999, Ramos-Franco et al. carried out experiments in which they induced the suppression of the first four α-helix transmembrane segments of the IP3R channels, leaving only segments 5 and 6 active. Even without the presence of the previous segments, the channel was formed with conductance, selectivity and properties that confirmed the proposed model. Furthermore, the homology of sequences between RyRs and IP3R channels in relation to pore formation suggests that studies with RyRs are important for the molecular determination of IP3R channels (Boehning et al., 2001).

3.2 - Proteins that remove calcium from the cytosol

The RS also contains *calsequestrin*, a protein with a high binding capacity for Ca^2 + (Ostwald & MacLennan, 1974) which is especially concentrated in the terminal cisternae (Meissner, 1975).

Two theories have been proposed to explain the mechanism of calcium ion release by the SR: 1) using the isolated SR of skeletal and cardiac muscles, it has been shown that this release is, within certain limits, a function of the concentration gradient of free calcium across the membrane of these vesicles (Fabiato & Fabiato, 1975). These data suggest that an increase in the activity of free calcium ions in the cytoplasmic medium can promote the release of calcium ions stored in the SR. This is the theory of "*calcium induced calcium release*" (Fabiato & Fabiato, 1977; 1983); 2) the other theory proposed is that the release of calcium by the SR is due to the depolarization of its membranes. In this theory, it is assumed that the inside of the vesicle is positively polarized in relation to its outside (Constantin et al., 1977). However, the latter has not been accepted.

Data aimed at evaluating the capacity of the SR to capture and store calcium indicates that the amount released is sufficient to trigger the process of muscle contraction and its reuptake is sufficient to induce muscle relaxation in a heart with a normal beat (Vaugham-Jones, 1986). This reuptake of calcium by the SR requires ATP and is blocked by calcium ionophores, suggesting the existence of a calcium pump located in the membranes of this organelle (Chapman, 1983; Vaughm-Jones, 1986).

4 - Excitation-Contraction Coupling Process (ECC)

Cardiac excitation-contraction coupling is the process from electrical excitation of the myocyte to the release of Ca^{2+} from the SR, which results in cardiac contraction (Bers, 2002). The second messenger calcium is essential in cardiac electrical activity and is the direct activator of the myofilaments that cause contraction (Bers, 2002). From a general viewpoint, the AEC process may suggest events that lead up to the end of the contractile movement, however, the correct conceptualization of the AEC process corresponds to all the events that begin with the depolarization of the membrane, whether in skeletal, smooth or cardiac muscle cells, up to the resulting release of Ca^2 + by the SR (Dulhunty, 2006).

Thus, the release of Ca^{2+} by the SR is of crucial importance for the AEC process. It has been reported that alterations in myocyte calcium concentration are the central cause of both contractile dysfunction and arrhythmias in pathophysiological conditions (Sordahl et al., 1973; Pogwizd et al., 2001).

In the heart muscle, the main mechanism involved in the release of Ca^{2+} through the SR is the ryanodine receptor or channel, RyR2 (Ogawa, 1994; Franzini-Armstrong & Protasi, 1997).

Numerous cellular structures are involved in the cardiac muscle AEC process, but as already mentioned, the events begin with membrane depolarization, which can be either electrical (voltage-dependent) or by chemical action (ligand-dependent). In cardiac myocytes, depolarization of the membrane triggers conformational changes that open voltage-dependent calcium channels (DHPRs) which enable the rapid influx of calcium (Katz, 1992; Clapham, 1995). Only a small amount of calcium that enters the cytosol from the extracellular space activates the adult myocardium, since most of this "activating" calcium accumulates just below the plasma membrane, adjacent to the calcium-releasing channels located in the sarcoplasmic reticulum (RyRs). This accumulation of Ca^2 + triggers the release of a greater amount of *calcium* from the sarcoplasmic reticulum, a mechanism identified as "*calcium-induced calcium* release" *(CICR*) (Clapham, 1995; Fabiato, 1983). The calcium present in the cytosol at this time is considered to be responsible for activating the contractile myofilaments (Clapham, 1995; Fabiato, 1983).

A second pathway has been proposed as a mechanism that contributes to the cardiac CICR mechanism, the *Na^+ / Ca^{2+} exchanger* (Bridge et al., 1990). This exchanger represents a mechanism by which Ca^{2+} movements through the sarcolemma are coupled to reciprocal Na+ movements; with a stoichiometry of $3Na^+$:$1Ca^{2+}$ (Brenner, 1993). The direction in which the exchange takes place depends fundamentally on the electrochemical gradient for Na+ and Ca^2 +. In the resting condition (diastole), the exchanger uses the electrochemical gradient that favors the entry of Na+ into the cell and the exit of Ca^2 +from it, thus acting, at that moment, as an exit route for intracellular Ca^2 +. In this condition, the exchanger is said to be acting in its direct or *forward* mode. However, when depolarization and the subsequent change in the electrochemical gradient occur, this also induces a change in the activity of the exchanger, reversing its direction, i.e. influx of Ca^{2+} and efflux of Na^+ . In this condition, the exchanger is said to be acting in its *reverse* mode. In this mode of action, the Na /Ca^{+2+} exchanger would be contributing to the *Ca^{2+} release* mechanism *induced by Ca^{2+}* . Obviously, by acting in its direct or forward mode, the Na /Ca^{+2+} exchanger would be contributing to the muscle relaxation process (Brenner, 1993; Sham et al., 1995).

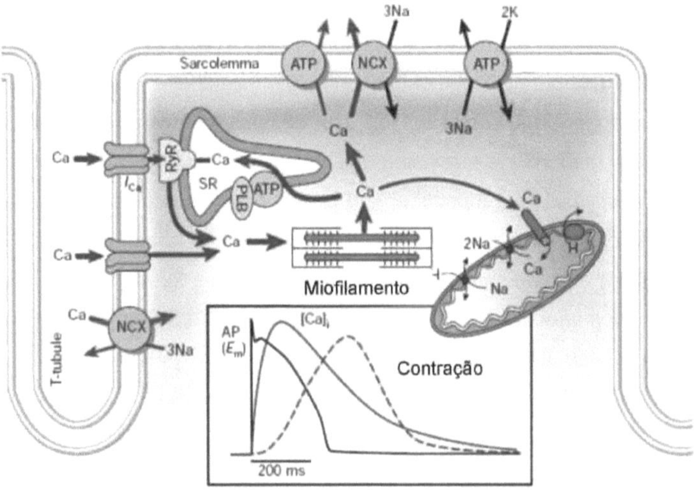

Figure 7 - *Calcium inducing calcium release* (CICR). Schematic drawing representing the AEC process by CICR. Slow Ca^{2+} channels allow the influx of this ion during membrane depolarization, increasing its cytosolic concentration, which in turn alters the activity of the Na^+/Ca^{2+} exchanger. When cytosolic calcium binds to RyRs, it induces the release of Ca^{2+} stored in the SR (Source: Bers, 2002).

Although, as a rule, the AEC process is described as we have done so far, in reality the variety and complexity of the structures involved in this process show us some substantial differences and complexities. Unlike ventricular myocytes, atrial myocytes have a poor tubular system (Hüser et al., 1996; Lipp et al., 1996a, b) and the coupling between the Ca channels^{2+} (DHPRs) located in the sarcolemma and the RyRs in the "junctional" portion of the SR occurs near the periphery of the cell (Lipp et al., 1990; Berlin, 1995; Carl et al., 1995; Hüster et al., 1996). Previous studies have indicated that the increase in the atrial concentration of Ca^{2+} occurs initially in the subsarcolemmal region, followed by a variable degree of propagation of the Ca^{2+} signal to the inner layers of the atrial myocyte (Lipp et al., 1990; Berlin, 1995; Hüster et al., 1996). Spatial modulation of Ca^{2+} signaling is critical for determining the cell's physiological activity. Different cell types have local and/or global Ca^{2+} signaling so that they can activate cellular processes differently (Bootman et al., 2001; Marchant & Parker, 2000). It has been shown that atrial cardiomyocytes alter the spatial properties of their Ca^{2+} signaling to modulate contractility. The ventricles are the main effectors of passive pumping and filling due to venous return pressure. However, atrial contraction contributes -20% of ventricular blood volume before they contract. The contribution of the atria becomes gradually more important with age or under stressful conditions (Mackenzie et al., 2004).

The AEC process is initiated at the periphery of the cell where DHPRs and RyRs are juxtaposed. In most cells, the response to Ca^{2+} is limited to the subsarcolemmal space. Since it doesn't occur in the neighboring areas of the contractile machinery, the Ca^{2+} signal from the junctional portion triggers only a modest jolt. The absence of inward propagation of the subsarcolemmal Ca^{2+} response is surprising given that atrial myocytes abundantly express non-junctional RyRs, which are functional and clearly have access to substantial stores of Ca^{2+} . The main mechanisms that act to limit the diffusion of Ca^{2+} are the Ca pump^{2+} - ATPase of the SR (SERCA) and the mitochondria (Mackenzie et al., 2004). The reason why SERCAs and mitochondria affect the intracellular propagation of the Ca^{2+} signal but not the initiation of the CICR process at the cell periphery is most likely the proximity between DHPRs and subsarcolemmal RyRs (- 20 nm) compared to the distance that exists between RyR clusters throughout the cell (-2 µm). However, pharmacological manipulation of RyR sensitivity through the CICR process, or by increasing the flow of the Ca^{2+} trigger, enables the subsarcolemmal Ca^{2+} signal to recruit central RyRs. This leads to an amplification of the global Ca^{2+} signal, resulting in a substantial increase in contraction force. Thus, researchers suggest that non-junctional RyRs are positioned in a strategic location and enable them to respond during the AEC process (Mackenzie et al., 2004). Under control conditions, the sensitivity of atrial myocytes to the CICR process is insufficient for them to be activated by the limited amount of Ca^{2+} that is not re-captured by SERCA and mitochondria. However, even a modest sensitization, for example using low concentrations of caffeine, makes them responsive. The same global modification of Ca^{2+} signaling is responsible for the positive inotropic effect of stimulation by endothelin-1 (ET-1); α-adrenergic agonist (phenylephrine) or β-adrenergic agonist (isoproterenol). Furthermore, in contribution to ventricular filling, atrial myocytes secrete hormones such as atrial natriuretic peptide (ANP). Of all these signals, the signal transduction pathway relating the response of atrial myocytes to α-adrenergic agonists is the least understood (Mackenzie et al., 2004).

Despite the strong evidence that calcium influx is essential for the excitation-contraction coupling process in cardiac muscle, some studies have suggested the existence of a voltage-dependent calcium release mechanism which does not require calcium influx (Ferrier & Howlett, 2001). However, some important studies have argued against this hypothesis, but so far nothing has been definitively proven (Wier & Balke, 1999; Piacentino et al., 2000).

Inositol (1,4,5)-triphosphate ($InsP_3$) can trigger the release of calcium from the SR of smooth muscle as well as from the endoplasmic reticulum of many cell types, via $InsP_3$ receptors.

There are InsP3 receptors in ventricular myocytes (basically isoform 2) (Perez et al., 1997; Lipp et al., 2000). Although a high concentration of InsP3 can release calcium in cardiac myocytes (particularly in atrial cells, which have a much higher expression of InsP3 receptors), the frequency and intensity of calcium release is much lower than that observed in the CICR mechanism and action potentials are known not to stimulate InsP3 production (Kentish et al., 1990). In addition, cardiac α-adrenergic and muscarinic agonists increase InsP3 production and contraction force (Brown & Jones, 1986; Pogglioli et al., 1986), however, this inotropic effect is mediated mainly by protein kinase C (PKC) rather than InsP3 (Endoh, 1996; Gambassi et al., 1998).

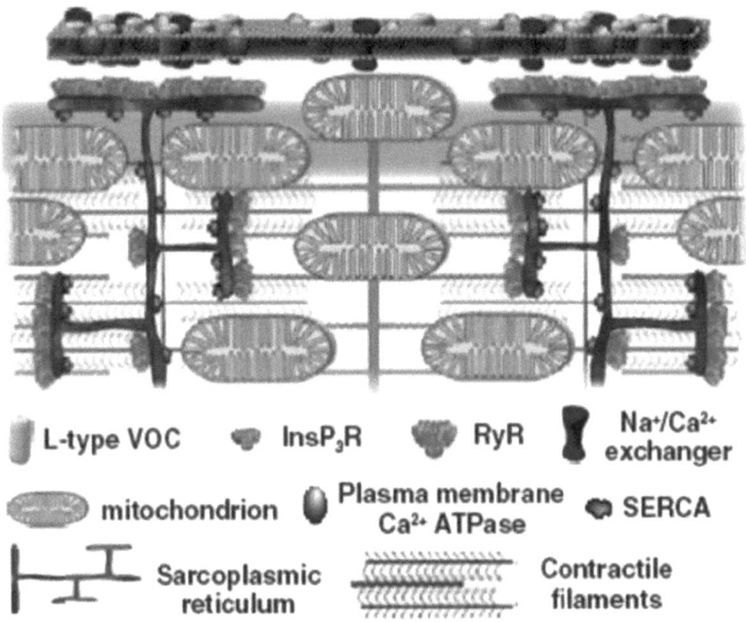

Figure 8 - Ca^2 + signaling machinery in atrial myocytes. The figure shows the specific arrangement of the components involved in Ca^2 + signaling in atrial myocytes, which contribute to the generation and modulation of Ca^{2+} signals in the atrial myocyte. The green band in the drawing delimits the zone in which the Ca^{2+} signal must cross in order to recruit the central RyRs and trigger contraction. (Source: Mackenzie et al., Journal of Cell Science, 117, 6327-6337, 2004).

4.1- Excitation-contraction coupling and the Inositol 1,4,5-triphosphate receptor pathway (IP3Rs)

Inositol 1, 4, 5-triphosphate (IP3) is an important second messenger in the regulation of intracellular Ca concentration^{2+} (Berridge and Irvine, 1984; 1989). The activation of certain

G protein-coupled receptors on the piasmatic membrane induces the hydrolysis of phosphatidylinositol 4,5-biphosphate (PIP2), with the consequent formation of IP3 and the subsequent release of Ca^{2+} from intracellular stores. In recent years, important studies have made significant progress towards a better understanding of the molecular basis for the release of Ca^{2+} via IP3. Among these advances, we can highlight the discovery of IP3Rs which are intracellular Ca^{2+} releasers (Moschella and Marks, 1993).

It has been shown that this process begins with the activation of receptors in the piasmatic membrane, most of which are made up of seven transmembrane domains and are coupled to a heterotrimeric complex of Gq proteins (, e). In the inactivated state, these Gq proteins are coupled to each other and to guanosine diphosphate (GDP). The binding of the drug to its specific receptor activates this protein complex, and the activated Gaq subunit then binds to guanosine triphosphate (GTP). The Gaq-GTP protein complex activates the enzyme phospholipase cβi (PL cβi), which is responsible for the hydrolysis of the membrane lipid phosphatidylinositol 4, 5-biphosphate (PIP2), and the consequent production of two second messengers, inositol 1,4,5-triphosphate (IP3) and 1,2-diacylglycerol (DAG). IP3 diffuses rapidly from the surface of the cell membrane to the IP3 receptors (IP3R) located on the SR membrane. The interaction of IP3 with the receptor induces a conformational change and the consequent release of Ca^2 + from the SR into the cytosol, which produces contraction. The important role of IP3 in the pharmacomechanical coupling of smooth muscle, and probably cardiac muscle, contrasts with the scarce role of this second messenger in the excitation-contraction coupling of skeletal muscle (Bezprozvanny and Ehrlich, 1995; Horowitz et al., 1996; Karaki et al., 1997).

The second compound hydrolyzed from PIP2 is DAG, made up of two fatty acids and glycerol, which remains associated with the membrane. DAG activates a serine/threonine kinase, *protein kinase C* (PKC). However, the activation of protein kinase C also requires the presence of a Ca signal^{2+} in the cytoplasm which promotes the recruitment of PKC to the membrane and therefore its interaction with DAG. As PKC activation occurs shortly after PLC activation, the simultaneous formation of IP3 and DAG ensures the synchronization of these two compounds. As a result, PKC phosphorylates a number of substrates, receptors, ion channels or enzymes, leading to a change in the functional state of the target cell (Kamoun, Lavoinne and Verneuil, 2006). Specifically in cardiac myocytes, the action of PKC is the phosphorylation of a membrane ion channel, selective for Ca^2 + ions, causing the transient flow of these ions into the cardiac cytosol (Endoh, 1996; Gambassi et al., 1998; DeSantiago et al., 1998).

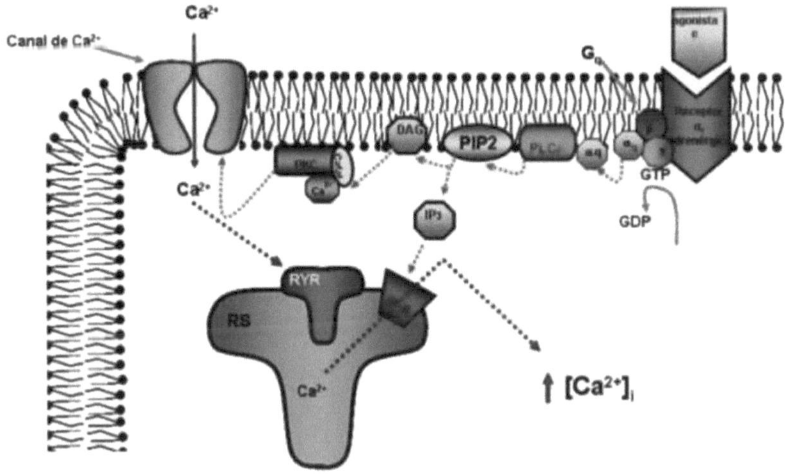

Figure 9 - AEC via α-adrenergic stimulation. Schematic drawing of the AEC process via α-adrenergic stimulation, identifying the sequence of events initiated by binding between the agonist and the receptor. This binding induces the activation of the Gs protein and the consequent cascade of events mediated by the PLCβ1 enzyme. This is followed by the activation of membrane calcium channels and the SR, finally resulting in the release of Ca^{2+} into the cytosol (Source: Bers, 2002).

5 - Essential Oils - Eugenol

Native to the Moluccas Islands, clove, along with black pepper, cinnamon and nutmeg, is one of the best-known spices in the world (Lavabre, 1997). Eugenol, the main component of clove oil, is a colorless or pale-colored liquid. It has a warm, spicy, characteristic fragrance; just like cloves. It is not very soluble in water and can be mixed with alcohol, chloroform, acetic acid and oils. Chemically, eugenol is a molecule of 2-methoxy-4-(2-propenyl) phenol; 4-allyl-2-methoxyphenol; allylguaiacol; eugenic acid; caryophyllic acid (Thompson et al., 1989). $C_{10}H_{12}O_2$; molecular weight= 164.2; d= 1.06.

As well as being the main component of clove oil, eugenol is also a constituent of the essential oils or extracts of many other plants, such as cinnamon and basil (Thompson et al., 1989; Wrigh et al., 1995). The Brazilian flora has a wide variety of plants that contain eugenol and/or its analogues in the composition of their essential oils. The *Croton* genus, for example, is very important in the Northeastern flora, not only because of its wide occurrence and dispersion in the cerrados, coastal tableland forests, rain forests and the Northeastern caatingas, but also because of its usefulness to the rural man (Ducke, 1959).

However, synthetic petroleum derivatives, obtained at lower prices, have forced a decline in the use of essential oils as a raw material for the international industry (Craveiro et al., 1981).

Eugenol, therefore, is not only found freely in nature, but can also be purchased commercially. It is therefore widely used in folk medicine in the preparation of teas and infusates which are considered to be beneficial in the treatment of some diseases such as gastrointestinal disorders and chronic diarrhea (Leal-Cardoso, 1994; Nishijima et al., 1998; Thompson, 1989).

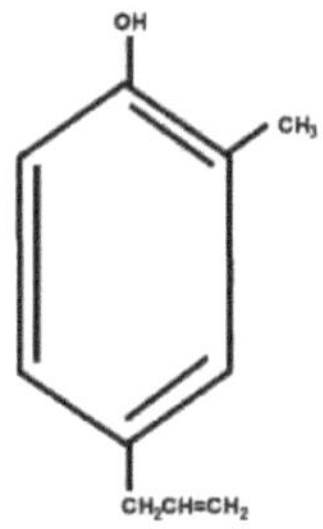

Three-dimensional structureFlat structure

Figura 10 - Chemical structure of the eugenol molecule.

In industry, it is used to accentuate the taste of sweets and drinks, and is also used in the manufacture of cosmetics (Leal-Cardoso, 1994). In dentistry, eugenol is excellent for relieving pain induced by dental pulpitis. It is also used in dental therapy, in the form of zinc-enolic paste, as a capping material for exposed pulps, as a temporary restorative for sensitive cavities and as a root canal filling material.

Because it is lipophilic, it is quickly absorbed (Escobar, 2002) and is able to cross biological membranes and reach intracellular targets such as mitochondria, where it inhibits the oxidation of NADH (reduced nicotinamide adenine dinucleotide), reducing ATP levels (Usta et al, 2002).

In a study of frog skeletal muscle, eugenol (0.12.5 mM) blocked the contraction induced by 80 mM K^+, an effect attributed to a decrease in the release of Ca^{2+} from intracellular stores. At higher concentrations (3-12 mM) eugenol promoted contraction independent of extracellular Ca^{2+} and blocked by procaine, suggesting that eugenol was releasing Ca^{2+} from the sarcoplasmic reticulum (Leal-Cardoso, et al., 1994).

The biological effects of eugenol depend on its concentration and the time of exposure to the tissue (Escobar, 2002). It has various pharmacological properties, such as antimicrobial, anti-inflammatory, antioxidant, immune response modulator, anticarcinogenic, cardiovascular, antinociceptive and local anesthetic. We can also highlight spasmolytic, antiseptic and muscle relaxant activity (Almeida, 2004), reduction of tongue edema induced by plant poisoning (Dip et al, 2004), anti-helminthic activity in ruminants (Pessoa et al, 2002).

Eugenol, at concentrations of 10 to 1000 μM, promoted vasodilation of the artery pre-contracted by norepinephrine, by high concentrations of K+ and by electrical stimulation, *in* both *in vivo* and *in vitro* experiments. These effects have been attributed to a blockade of calcium channels and inhibition of the contractile machinery (Hume, 1983; Nishijima et al, 1999). Intravenous treatment of rats with eugenol induces arterial hypotension from a dose of 1mg/Kg in both anesthetized and awake rats. *In vitro*, eugenol showed vasodilator activity in a mesenteric bed precontracted by 60 mM K^+ (Lahlou et al, 2004).

Sensch et al. (2000) demonstrated that eugenol exerted negative inotropic effects on cardiac muscle similar to those observed when calcium channels are blocked with nifedipine. This negative inotropic effect may be the result of blocking membrane calcium channels. It has also been suggested that the reduction in contractile protein activation observed at basal tension may be the result of the depletion of intracellular Ca^{2+} reservoirs during the resting period, which may be the result of blocking membrane Ca^{2+} channels (Wendt-Gallitelli, 1985).

In studies carried out by Damiani et al (2004), it was observed that eugenol depresses myocardial contractility in a concentration-dependent manner and also reduces the time parameter of isometric contractions. The negative inotropic effect of eugenol was reversed by ISO and Ca^{2+} , and this effect was not accompanied by a reduction in relative potentiation, suggesting that there is no interference with SR activity. The depressant action of eugenol on contraction without interfering with SR function reinforces the hypothesis of reduced Ca influx^{2+} . The depressant effect of eugenol on tetanic contractions suggests that its action is supposedly on contractile proteins, but no reduction in myosin ATPase activity was observed. As a reduction in the influx of Ca^{2+} was observed, it is suggested that this oil acts by blocking membrane calcium channels, which seems to be the reason why it has a negative inotropic effect.

In studies carried out on intact atrial muscle, Olivoto (2008) observed that the effects of eugenol on the AEC are concentration-dependent and that at high concentrations, this compound is capable of both abolishing muscle jerking and inducing an increase in resting tension.

RyR blockers were unable to alter the development of resting tension induced by eugenol, indicating that the pathway by which this compound was acting was independent of RyRs. In intact atrial muscle, the IP3R blocker (2-APB) was able to partially reduce eugenol's ability to increase force, suggesting that this oil may be acting via IP3Rs (Olivoto, 2008).

In saponin-permeabilized fibers, it was observed that Heparin, an IP3Rs blocker, was able to reduce eugenol's ability to increase strength. As observed with 2-APB, the blockade was partial, indicating that part of eugenol's action is through IP3Rs, however, suggesting that eugenol could also be acting through a second route of action (Olivoto, 2008).

It was also observed, in permeabilized fibres, that the force induced by eugenol is an active state of cyclization of the cross-bridges, since BDM was able to reverse it, and that its action is not directly on the contractile proteins since its action was reversed when the preparations were permeabilized by Triton X-100. The same was not observed in a solution with a high concentration of EGTA (Ca chelator^{2+}), under which conditions eugenol was able to induce an increase in strength. The results therefore suggest that eugenol is probably an agonist of IP3Rs, thus inducing the release of Ca^2 + from the SR. However, eugenol may also have a second, unknown pathway of action, independent of Ca^2 +, and possibly with a greater capacity to activate contractile proteins than that observed by the IP3Rs pathway (Olivoto, 2008).

OBJECTIVES

GENERAL OBJECTIVE

This study aims to evaluate the effects of eugenol on the resting tension of rat atrial muscle trabeculae in the absence of calcium.

SPECIFIC OBJECTIVES

a) To assess whether eugenol acts directly on the contractile protein complex.

b) To determine whether eugenol is capable of inducing an increase in resting tension in solutions lacking calcium.

c) Identify the possible routes of action of eugenol.

MATERIALS AND METHODS

1- Experimental Procedures

For this study, 280 male Wistar rats of approximately 3 months of age, weighing between 250 - 300g and supplied by the Bioterium of the Biological Sciences Sector of the Federal University of Paranà, were used. The animals were kept in cages under controlled temperature conditions and a 12-hour light/dark cycle, with free access to food and water. The care and use of the animals complied with the "Rules for the Use and Handling of Experimental Animals" of the Federal University of Paranà and in accordance with the recommendations of the Helsing Convention.

The experimental procedures were approved by the Ethics Committee in Animal Experimentation of the Federal University of Paranà, Biological Sciences Sector, certificate nº 231, process 23075.021110/2007-11, approved on June 29, 2007. The experiments were carried out from August 2007 to November 2011.

The rats were anesthetized by inhalation of ethyl ether and exsanguinated immediately after anesthesia. The hearts were removed quickly after thoracotomy. After removal, the hearts were perfused with Ringer's solution through the aortic stump to allow proper selection and dissection of the atria. The time between thoracotomy, removal of the hearts and dissection of the atria did not exceed 1 minute. After dissecting the atria, trabeculae were dissected to be used in the experiments. The trabeculae were dissected from both the right and left atria. Eight trabeculae were dissected from each pair of atria (Figure 11).

The force developed (F) and/or the Resting Tension was assessed using a force transducer (Scientific Instruments, GMBH, Heidelberg, Germany) (Figure 12) and recorded on a polygraph (Beckman 5525; Type RM Dynograph® Recorder). Before each experiment, the force transducer calibration curve was obtained with known masses. The length and width of the trabeculae were also measured (assuming that the trabeculae were cylindrical), using a graticule positioned in the stereo microscope eyepiece. The cross-sectional area of the preparations was calculated using the following formula:

Area = mass / (length x density)

Thus, the force produced by the atrial trabeculae was normalized by their cross-sectional area and expressed in $mN/mm.^2$

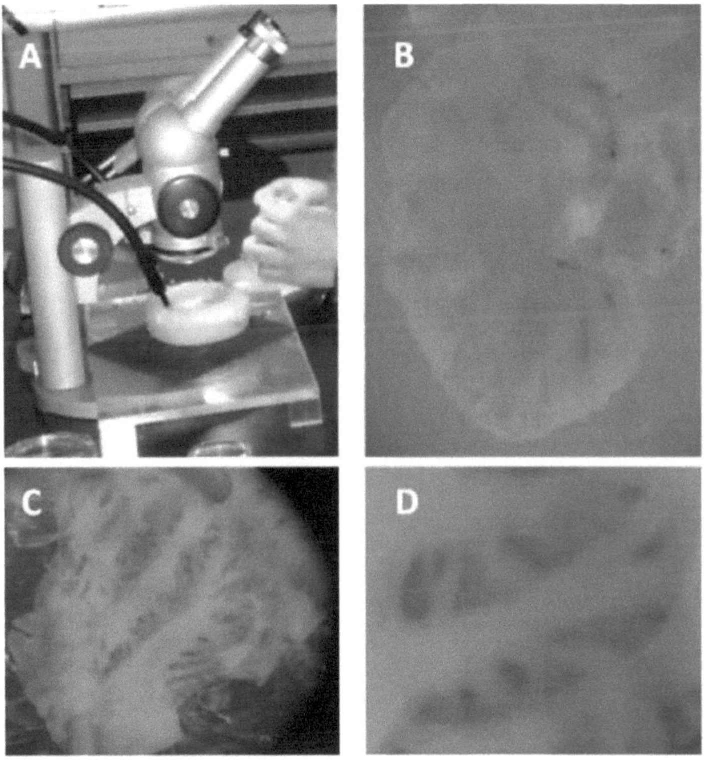

Figura 11 - Photos of the experimental procedures for dissection. A - photo of the stereo

Zeiss microscope used to dissect the atria and trabeculae; B - photo of an atrium after it has been dissected; C - internal view of an atrium, which shows the arrangement of the trabeculae, making it possible to identify the trabeculae to be dissected; D - photo of a single trabecula that was dissected for use in the experiments. Source: Olivoto.

2 - Procedures for Experiments with Intact Trabeculae

After dissection, the trabeculae were mounted in a vat with a volume of 2.5 ml and continuously perfused with experimental solution. To avoid the possibility of hypoxia, the experiments were carried out at a low temperature (28+1° C) as previously described (Vassallo et al., 1994) and continuously gassed with 95% oxygen (O_2) and 5% carbon dioxide (CO_2). The trabeculae were assembled and stretched to a maximum length L_{max} (muscle length at which the active tension is maximum) and stimulated using rectangular pulses (10 to 15 V; 12 milliseconds in duration) through a pair of platinum electrodes positioned along the entire length of the muscle. The standard stimulation frequency was 0.5 Hz (constant condition). The preparations were kept for a stabilization period of 20 minutes to allow them to adapt to the new environmental conditions and then the experimental protocols were started (Figure 12).

For the experiments with intact trabeculae submitted to indirect electrical stimulation, we used the following solutions: Normal Ringer's Solution, composed of (in mM): 110 NaCl; 4.0 KCl; 2.0 $MgCl_2 6H_2O$; 2.0 $CaCl_2 2H_2O$; 10.0 Trisma; 11.0 Glucose; pH 7.4; and zero-Ca Ringer's Solution^{2+} , composed of (in mM): 112 NaCl; 4.0 KCl; 2.0 $MgCl_2 6H_2O$; 10.0 Trisma; 11.0 Glucose; pH 7.4. As eugenol is an oil and not very soluble in water, we kept both the stock solution and the bath solution constantly stirring to ensure its solubility.

2.1 - Effects of eugenol on the resting tension of atrial muscle trabeculae: Dose-Response Curve

In order to assess the effect of eugenol on the resting tension (F) of the atrial muscle, the following protocol was drawn up: After a stabilization period of 20 minutes and the preparation kept under standard stimulation, cumulative concentrations of eugenol (1, 3, 5, 7 and 10 mM) were added to the bath solution (normal ringer's), with a time interval of 15 minutes between each concentration. This period was sufficient for the resting tension to reach steady state.

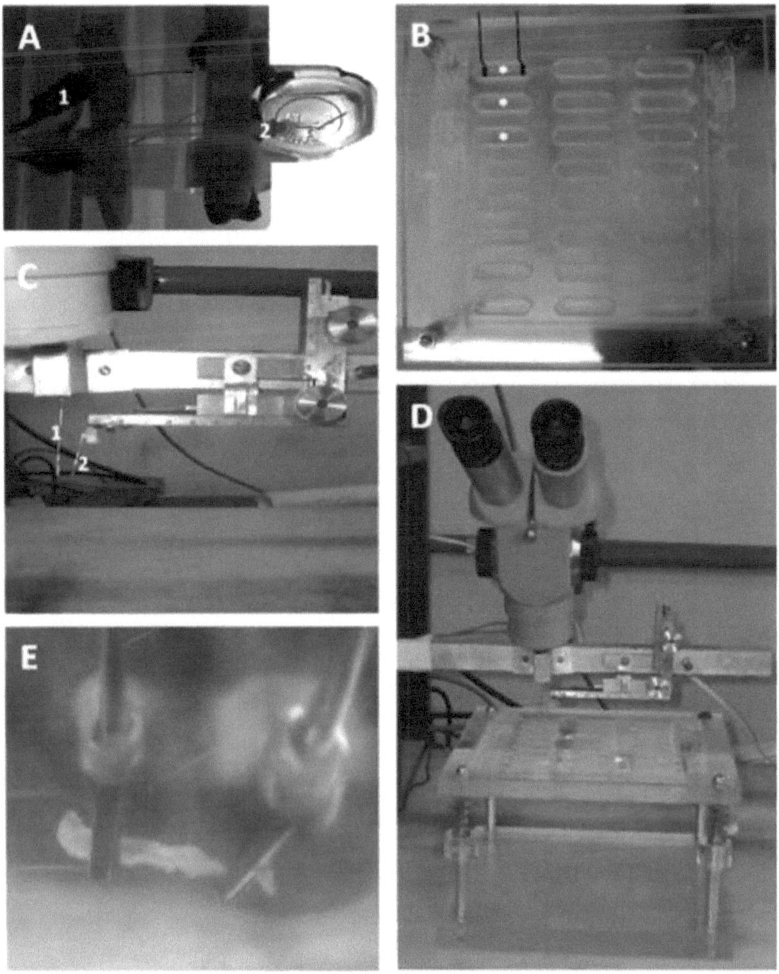

Figure 12 - Photos of the equipment used to carry out the experiments; A.1 - Indirect electrical stimulation system; A.2 - gasification system; B - photo of the acrylic platform containing the chambers (*) where the experimental solutions were added; C.1 - force transducer; C.2 - D - photo of the complete system for carrying out the experiments, containing the stereo microscope, force transducer and acrylic platform; E - trabecula mounted between the force transducer and the mobile mechanical arm. Source: Olivoto.

2.2 - Effects of eugenol on intracellular and extracellular calcium mobilization in normal ringer and zero calcium ringer.

Calcium-induced calcium release (CICR) represents an important stage in the cardiac muscle AEC process, and control over cardiac contractility can be exerted through changes in the extracellular concentration of calcium $[Ca^{2+}]_o$ (Yue et al., 1986). Procaine is a local anesthetic, capable of preventing the generation and conduction of nerve impulses through its direct inhibitory interaction with voltage-dependent sodium channels, thus blocking ion influx. It is also capable of exerting an inhibitory effect on RyRs by interfering with the calcium release mechanism of the SR (Ostfeld et al., 1997; Rang et al., 2001).

To determine whether or not eugenol was acting on the main Ca releasing channels^{2+} of the RS (RyRs) (Normal Ringer's solution) and/or whether the concentration of extracellular calcium, $[Ca^{2+}]_o$, could alter the results (zero-Ca Ringer's solution^{2+}), we developed a protocol in which, after standard indirect electrical stimulation, the atrial muscle was treated for 10 minutes with procaine (10 mM). In the experiments with zero-Ca Ringer's Solution^{2+} , electrical stimulation was carried out in Normal Ringer's Solution, after which the solution was replaced by zero-Ca Ringer's Solution^{2+} After this period of incubation, eugenol (5 mM) was added to the bath solution (normal ringer's and zero-Ca ringer's^{2+}).

Despite the fact that procaine proved incapable of inhibiting the increase in resting tension induced by eugenol and in order to ensure that eugenol may or may not be acting through RyRs, we developed another protocol. We designed an experimental protocol using the RyR blocker, ryanodine, an alkaloid found naturally in the stem and roots of the plant *Ryania Speciosa*. This compound has a high affinity for RyRs, with an inhibitory action. The experiments were carried out in normal ringer's solution and zero-Ca ringer's solution^{2+} . In the experiments with zero-Ca ringer solution^{2+} , electrical stimulation was carried out in normal ringer solution, and after the incubation period the solution was replaced with zero-Ca ringer solution^{2+} After a period of stabilization of the preparation under indirect electrical stimulation, we added ryanodine (100 μM) to the bath solution, leaving it there for 10 minutes. After the incubation period, eugenol (5 mM) was added to the bath solution.

In order to control all the variables involved in the release of calcium from the RS, we designed a protocol in which we used two RyR blockers together, procaine (10 mM) and ryanodine (100 μM). The protocol was carried out in normal ringer's solution and zero-Ca ringer's solution^{2+} . The initial experimental procedures were the same as those described in the previous paragraphs. After a 10-minute incubation period, eugenol (5 mM) was added to the solution.

2.3 - Effects of eugenol on contractile proteins

To determine whether eugenol's action could be directly on contractile proteins, we developed a protocol using BDM (2,3 butanedione monoxide) (30 mM), which acts by significantly reducing the maximum force-generating capacity of the contractile system because it is capable of stabilizing the ADP+PI complex, resulting in the impossibility of binding new ATP at the end of the cross-bridge cycle (Li et all, 1984; Stephenson & West, 1989; Higuchi & Takemori, 1989).

After standard stimulation of the trabeculae in normal ringer's solution, we added procaine (10 mM) to prevent the possible release of calcium from the SR, and left them in these conditions for 10 minutes. After this time, eugenol (5 mM) was added. After reaching the maximum resting tension, BDM (30 mM) was added to the bath solution. The preparation was kept under these experimental conditions until the BDM-induced relaxation stabilized (20 to 30 minutes).

In a new protocol, we repeated the initial experimental procedures in normal ringer's solution (described in the previous paragraph). After the incubation period with procaine (30 mM), the bath solution was replaced with zero-Ca ringer solution^{2+} , plus procaine (30 mM) and eugenol (5 mM). After reaching maximum resting tension development, BDM (30 mM) was added to the bath solution. The preparation was kept under these experimental conditions until the BDM-induced relaxation stabilized (20 to 30 minutes).

In both protocols, after the action of the BDM, we washed the preparation in normal ringer solution 5 times. After washing, we transferred the preparation to eugenol normal ringer solution (5 mM).

2.4 - Effect of Eugenol after depletion of intracellular calcium stores by Caffeine

The possibility of depleting the Ca^{2+} stored in the SR was observed by Bradley et al (2002) in smooth muscle cells. The SR was depleted of Ca^{2+} by the action of caffeine (10 mM) applied sequentially at intervals of 1 to 2 minutes. On the other hand, caffeine (30 mM) was used in experiments with crustacean muscles, resulting in the activation of RyRs and the consequent release of stored Ca^{2} + (Lea, J. T., 1996). Caffeine has the property of binding effectively to RyRs, in other words, as long as there is caffeine in the solution these receptors remain activated, which enables the constant release of Ca^{2+} from the SR (Bradley et all, 2002).

A second pathway has been proposed as a mechanism contributing to cardiac CICR, the *Na^{+} / Ca^{2+} exchanger* (Bridge et al., 1990). These exchangers are responsible for the movement of Ca^{2} + through the sarcolemma, which is coupled to reciprocal movements of Na+ (Brenner, 1993).

We designed this protocol to test two hypotheses: the first relates to the possibility of eugenol acting cooperatively through these exchangers; the second is whether eugenol was able to increase the resting tension even after the RS was depleted by caffeine. The protocol was carried out with intact trabeculae.

To carry out this protocol, we replaced normal ringer's solution with zero-Ca ringer's solution^{2+} and zero-Ca ringer's solution /Na^{2++} , both with caffeine (30 mM).

In zero-Ca^{2} $_{+/Na+}$ ringer solution plus caffeine, it was possible to inhibit the activity of the exchanger in both the direct and indirect directions, and to maintain the constant release of Ca^{2} + from the SR due to the action of caffeine (30 mM), thus enabling the establishment of a plateau in force production. After obtaining maximum force in caffeine, observed by the generation of a plateau, we transferred the preparation to zero-Ca^{2+} ringer solution, plus caffeine (30 mM). In these experimental conditions, we kept the exchanger inhibited in the indirect direction, but allowed it to activate in the direct direction, facilitating the efflux of Ca^{2+} into the bath solution. These procedures were repeated 2 to 5 times until no further response was observed in the presence of caffeine, indicating that the SR had been exhausted. The preparation was then transferred to zero-Ca Ringer's solution^{2+} plus caffeine (30 mM) and eugenol (5 mM). A new set of trabeculae was used to determine the control eugenol.

3 - Procedures for Experiments with Trabeculae

Permeabilized with Saponin and/or Triton X-100

In order to evaluate the possible effects of eugenol on the muscle contractile apparatus, on the sarcoplasmic reticulum and on the sequence of chemical reactions that lead to the phosphorylation and activation of muscle contractile proteins, we carried out experiments using the *skinned* fiber technique. This preparation has proved particularly useful because it allows us to keep the composition of the solution that will bathe the intracellular organelles and the contractile system under strict control. With the use of a permeabilizing solution containing saponin, 30 µg/ml (Sigma Chemical), a glycoside that acts by interacting with cholesterol molecules present in the lipid bilayer of the sarcolemma and the transverse tubular system, it is possible to selectively remove this planar chain lipid and promote perforations in the membrane (Launikonis and Stephenson, 1997). These structures are selectively permeabilized, preserving, in addition to the proteins of the contractile system, the RS and its membrane structures. Using this method, data was obtained on the behavior of calcium through the permeabilized membrane of this organelle. To study the effects of eugenol on the action of calcium exclusively on the actin-myosin contractile apparatus, a purified permeabilizing solution containing 1% vol/vol Triton X-100 (Boeringer) was used. Triton X-100 is a non-ionic detergent with the ability to solubilize lipid structures such as the sarcolemma and the membranes of all intracellular organelles, leaving only the contractile apparatus intact and preserved in the intracellular environment (Godt et al., 1991; Lamb et al., 1994), 1991; Lamb et al., 1994).

The atrial muscles were prepared as described above. Once dissected, the atria were transferred to a *Petri* dish containing Solution R (Table 1), in which the trabeculae were dissected. The trabeculae were then transferred and kept for 20 minutes in a chamber containing the same Solution R plus saponin (30 μg/mL) or for 10 minutes in Solution R plus 1% (v/v) Triton X-100, a non-ionic detergent. After permeabilization by saponin or Triton X-100, the trabeculae were transferred to another chamber containing Solution R and kept there until the start of the experiments.

To calculate the composition of the solutions, the turbo language computer program was used, which describes the multiple equilibrium constants of ions in solution using previously published association constant values (Godt & Lindley, 1982; Andrews et al., 1991). The compositions of the solutions are shown in Table 1.

Table 1- Composition of the solutions used in the experiments with fibers permeabilized by Saponin and by T riton X-100.

	Solution R	Solution L	Solution E	Solution A
Mg^2 + (mmol/L)	1	1	1	0,1
MgATP (mmol/L)	1	1	1	1
NaCP (mmol/L)	15	15	15	15
EGTA (mmol/L)	5	5	0,5	5
BES	50	50	50	50
pCa (-log[Ca^{2+}])	>8,5	6,4	>8,5	4,0
pH	7	7	7	7
Ionic strength	200	200	200	200
Temperature (° C)	22	22	22	22

Note: NaCP, Na2phosphocreatine; BES, N,N-Bis (2-hydroxyethyl)-2-aminoethanesulfonic acid; pCa = -log[Ca^{2+}]o; and pH = -log [H].$^+$

The solutions described in Table 1 have concentrations of 0.5 mM and 5 mM EGTA (ethylene glycol-bis (β-aminoethyl ether)-N,N,N',N'-tetraacetic acid) which is an agent with the ability to chelate Ca ions^2 + from the solution (Godt & Lindley, 1982; Andrews et al., 1991).

Solution R (pCa 8.5) (pCa = -log[Ca^{2+}]) is called a relaxing solution because it has a concentration of Ca^2 + ions lower than $10^{-8.5}$ M and a high concentration of EGTA, which allows it to chelate all the Ca^2 + that may still be in the bath solution. It has been described in the literature that caffeine, a known RyR agonist, at a concentration of 30 mM, is capable of inducing the total release of Ca^{2+} stores from the SR (Endo, 1975).

Solution A (pCa 4.0) has a sufficient concentration of Ca^2 + ions for maximum activation of the contractile apparatus and consequently the maximum capacity for force development by the trabeculae. For quantification purposes, the force developed by the trabeculae when in pCa 4.0 was determined as the maximum force production standard.

Solution L (pCa 6.4) has a concentration of Ca ions^{2} + sufficient to induce, by active transport, the uptake of Ca^{2+} into the SR. We can therefore say that solution L has the property of recharging the RS with Ca^{2} + ions. It is therefore possible, under these experimental conditions, to subject the same trabeculae to different solutions, with different agents, thus enabling a direct comparative analysis between the various treatments.

Solution E (pCa 8.5) has a Ca^{2} + ion concentration of less than 10^{-8} M. As the concentration of EGTA is low (0.5 mM), its capacity to buffer Ca^{2+} ions is small. Therefore, when Ca^{2} + ions are released from the SR, this ion (because it is not buffered) will activate the contractile system, which will develop tension, and this will provide an indication of the amount of Ca^{2+} released from the SR.

As eugenol is an oil and not very soluble in water, the stock solution was vigorously stirred before being added to the bath solution to ensure its solubility.

3.1 - Effects of eugenol on force production in atrial trabeculae permeabilized by saponin and Triton X-100

To test the effects of eugenol on intracellular Ca^{2} + stores and on the contractile apparatus, we developed the following protocol: After being mounted between a force transducer and a movable arm, the trabeculae were stretched 20% in relation to their size in the relaxation condition, and kept for a few minutes in R solution.

From solution R, the trabeculae were transferred to solution A, where they remained long enough to determine the maximum force induced by the high concentration of Ca^{2} + ions in the solution. This time was sufficient to activate the entire contractile apparatus.

The trabeculae were then transferred back to the relaxing R solution and kept there for a few minutes. These minutes were enough time for the preparation to completely relax as a result of the action of EGTA.

The trabecula was then transferred for a period of 2 minutes to Solution L. This allowed the RS to be recharged with Ca^{2} + ions. Data from our laboratory indicates that this time is sufficient to obtain the RS's maximum charging capacity (data not shown).

The trabeculae were then transferred and kept in Solution E for 30 seconds. From Solution E, the preparation was transferred to another chamber containing Solution E plus caffeine (30 mM). The preparation was kept in this condition long enough to activate the contractile apparatus and produce force as a result of the Ca^{2+} ions released from the SR by the action of Caffeine.

The preparation was then transferred to Solution R so that the Ca^{2+} ions released by caffeine could be chelated and muscle relaxation could occur. Once relaxation had been achieved, the preparation was transferred to Solution E. The preparation was then transferred to a new chamber containing Solution R plus Eugenol (5 mM). The preparation was kept in this chamber until the resting tension reached a stabilization plateau. Triton X-100 1% was then added to the solution.

3.2 - Effects of ruthenium red (RR) on the action of eugenol in the AEC process of saponin-permeabilized fibers

We developed a protocol in which the preparation was treated with ruthenium red (10 μM). Ruthenium red (Ammoniated Ruthenium Oxycloride) is an agent well described in the literature and used as an inhibitor of mitochondrial Ca^{2+} transport (Gunter, 1990) as well as an inhibitor of Ca^{2+} channels in smooth muscle, striated muscle and cardiac SR (Vites and Pappano, 1994; Chamberlain et al., 1984). The atrial trabeculae were prepared and mounted as described above, only this time, after passing through the chambers containing solution A and the chambers containing solution R, the trabeculae were transferred and incubated for 20 minutes in *chambers* containing solution E plus ruthenium red (10 μM).

The preparation was then transferred to a new chamber containing Solution R containing ruthenium red (10 μM) and eugenol (5 mM). The preparation was kept under these conditions until the resting tension reached a stabilization plateau.

3.3 - Effects of BDM on the action of eugenol in the AEC process of saponin-permeabilized fibers

We devised a protocol in which we used BDM (2,3 butanedione monoxide), a well-described blocker of the cross-bridge cycle (Li et all, 1984; Stephenson & West, 1989; Higuchi & Takemori, 1989; Herrmann et all, 1992). The atrial trabeculae were prepared and mounted as described above. After obtaining maximum force in solution A, and total relaxation in solution R, we transferred the trabeculae to a chamber containing solution R and Eugenol (5 mM). After the development of the maximum resting tension induced by the action of eugenol, we transferred the trabeculae to a chamber containing solution R plus eugenol (5 mM) and BDM (10 mM). The preparation was kept in these conditions until stabilization was achieved.

Subsequently, the same protocol was repeated, but in the presence of BDM (30 mM).

After stabilization was achieved in this experimental phase, the trabeculae were washed 5 times and subjected to R solution plus eugenol (5 mM) to obtain control eugenol data.

3.4 - Relationship between different concentrations of calcium ions in the bath solution and the resting tension induced by Eugenol

The technique of permeabilized fibers allows us to use solutions with strict control of composition and concentration, so we carried out a protocol in which we used different concentrations of Ca^{2+} in solution, starting at $10^{-8,5}$ and ending at $10^{-4,0}$. The aim of this protocol was to determine whether eugenol acts directly on the contractile machinery, whether it acts by sensitizing contractile proteins to Ca^{2+}, or whether it acts by promoting the extraction of Troponin I.

For the experiments, we used atrial trabeculae permeabilized with 1% (Vol/Vol) T riton X-100, thus keeping all the cell proteins intact, including the contractile proteins. The protocol was carried out in the presence and absence of eugenol (5 mM).

3.5 - Effects of eugenol on the AEC process before and after atrial fibers are permeabilized with Triton X-100

The permeabilization technique allows us to use solutions that directly bathe the cell organelles. For this reason, we developed a protocol in which we subjected the trabeculae to permeabilization by two techniques, i.e. the same trabeculae were permeabilized by saponin and by Triton X-100.

T riton X-100, as already described, is capable of solubilizing lipid structures in both the plasma membrane and the membrane of intracellular organelles. EGTA, also already described, acts by chelating free cellular Ca^{2+}. We designed this protocol to answer the following questions: a) EGTA is capable of chelating the Ca^{2+} released from the SR by the action of caffeine; b) even at high concentrations of EGTA, eugenol is capable of inducing an increase in resting tension; c) after the trabeculae are permeabilized by Triton X-100, eugenol is still able to induce an increase in resting tension; and d) permeabilization by Triton X-100 may somehow alter the ability of contractile proteins to develop strength in the presence of Ca^{2+}.

The experimental procedures were as follows: the trabeculae were permeabilized by saponin (30 µg/mL - 20 minutes), after permeabilization we obtained the maximum force in solution A, after relaxation in solution R, they were transferred to solution L, for loading, prepared in solution E and then transferred to solution R (high EGTA) plus caffeine (30 mM). They were then transferred to solution R plus eugenol (5 mM) and, after stabilizing in this solution, transferred to solution R plus Triton X-100 (1% - Vol/Vol - for 10 minutes). After permeabilization with Triton, they were transferred to R solution plus eugenol (5 mM). To finish the protocol, the trabeculae were transferred to solution A and kept in these conditions until they reached the stabilization plateau.

3.6 - Effects of saponin exposure time on calcium- and caffeine-induced force production capacity and Eugenol-induced resting tension elevation

We designed a protocol in which the trabeculae were permeabilized at different times: intact trabeculae (time zero), trabeculae permeabilized (time 20, 40, 60 and 80 minutes) by saponin. The aim of this protocol was to determine whether permeabilization was necessary to obtain a response in solution A; whether permeabilization by saponin resulted in any alteration in the response to caffeine, i.e. whether saponin could be acting on the cell organelles; and to determine whether the increase in resting tension induced by eugenol changed over the permeabilization time.

The trabeculae were exposed to solution A to determine the integrity of the contractile proteins; solution E plus caffeine (30 mM) to determine the integrity of the RS; and solution R plus eugenol (5 mM) to determine the ability of eugenol to increase the resting tension over the time of exposure to saponin.

3.7 - Effects of 1,4-Dithiothreitol (DTT) on the action of Eugenol

We devised a protocol using 1,4-Dithiothreitol (DTT) (100 µM), a reducing agent, which is able to protect proteins in solution from degradation and is not able to induce force production (Posterino & Lamb, 1996). DTT was added to the R solution during permeabilization (20 minutes). For the experiments we used Solution R plus DTT, to determine whether this compound was capable of activating the contractile machinery or capable of inducing force production in some other way; and Solution R plus Eugenol, to determine whether the action of DTT could in any way alter the ability of eugenol (5 mM) to induce an increase in resting tension. A new set of trabeculae, permeabilized by saponin (20 minutes), was used to determine the control eugenol.

3.8 - Effects of Colchicine (colch) on the action of Eugenol

To complement the previous protocol, we carried out a set of experiments using Colchicine (30 µM). Colchicine is a microtubule-disrupting agent. Microtubules are the structural support for the attachment and trafficking of proteins and cytoplasmic enzymes. The aim of this protocol was to determine whether the possible pathway by which eugenol acts is associated with microtubules. Thus, if the agent through which eugenol acts is a microtubule-associated protein (MAP), disorganization will possibly potentiate the action of eugenol. With the disorganization of the microtubule, the concentration of the possible agent by which eugenol acts would increase significantly, resulting in the potentiation of the action of this essential oil, even in the presence of a high concentration of EGTA. After the saponin permeabilization period, the trabeculae were transferred to solution A to determine the maximum force. They were then relaxed in solution R, then transferred to solution L to load the RS, and then transferred to solution E plus caffeine (30 mM). They were then transferred to solution R plus colchicine (30 µM), where they remained for 20 minutes. After the incubation period, they were transferred to R solution plus eugenol (5 mM) and colchicine (30 µM). A new set of trabeculae, permeabilized by saponin (20 minutes), was used to determine the control eugenol.

3.9 - Experiments with the product of the permeabilization of the trabeculae and their effect on Eugenol

As a complement to the previous protocols, we decided to test the viability of reincorporating the product of the permeabilization of the trabeculae within 80 minutes into trabeculae permeabilized with Triton X-100. For the experiments, we produced two experimental solutions called "SUCO" and "SUCO+COLCH". We permeabilized approximately 60 mg of atrial tissue for 80 minutes with saponin in solution R. The product of this permeabilization, i.e. what was lost from the tissue to the bath solution, was called SUCO. The same amount of atrial tissue, i.e. 60 mg, was permeabilized for 80 minutes with saponin plus colchicine (30 µM) in R solution; the product of this permeabilization was called SUCO+COLCH.

The trabeculae used for the experiments were permeabilized with 1% (vol/vol) Triton X-100 for 10 minutes. This protocol was designed to test the following hypotheses: after permeabilization by Triton X-100 we no longer observed the caffeine response due to solubilization of the SR membrane; after permeabilization by Triton X-100 eugenol loses its ability to induce an increase in resting tension; the pathway by which eugenol acts is cytosolic and soluble within 80 minutes of permeabilization by saponin; and permeabilization by saponin in the presence of a mattress potentiates the effects of eugenol.

The experimental procedures were as follows: solution A to obtain the maximum force and determine the integrity of the contractile machinery; transfer, after loading in solution L, to solution E plus caffeine (30 mM) to determine whether the RS was solubilized by the action of Triton X-100; transfer to solution R plus eugenol (5 mM) to determine whether this oil acts directly on the contractile machinery; transfer to a chamber containing SUCO, remaining in this condition until stabilized, and then transferred to SUCO plus eugenol (5 mM). A new set of trabeculae, also permeabilized with Triton X-100, were subjected to the initial steps described above. After passing through R solution plus eugenol (5 mM), they were transferred to SUCO+COLCH solution, remaining in this condition until stabilization, then transferred to SUCO+COLCH solution plus eugenol (5 mM). A new set of trabeculae, permeabilized by saponin (20 minutes), was used to determine the control eugenol.

4 - Possible routes of action of Eugenol

4.1 - Possible involvement of the p38 MAPk pathway

In studies carried out by Vahebi et all (2006), it was observed that acute activation of p38 MAPk results in a decrease in the ability to develop tension/strength in cardiac myocytes. To test this hypothesis, we used sodium arsenite in our experiments, which is a p38 MAPk agonist.

a) Dose response curve to sodium arsenite

To determine the ideal concentration of sodium arsenite for our experiments, we carried out a dose response curve on intact trabeculae. The trabeculae were subjected to indirect electrical stimulation (as described in section 2 of this chapter), after a period of stabilization, we added eugenol (5 mM) to normal ringer's solution, and after the development of maximum resting tension, we cumulatively added sodium arsenite at concentrations of 1.3, 5, 7 and 10 mM. This protocol allows us to determine the average concentration of sodium arsenite to be used in the other protocols.

b) Recovery after exposure to sodium arsenite

As sodium arsenite is a toxic agent and could in some way alter the activity of contractile proteins, we devised a protocol to test the viability of the trabeculae after exposure to this agent.

The trabeculae were electrically stimulated and after stabilizing the preparation under these experimental conditions, sodium arsenite (7 mM) was added to the bath solution (normal ringer). After stabilization, the preparation was washed 5 times and again subjected to electrical stimulation in normal ringer solution to determine whether there was any deleterious action of the arsenite in the preparation. After stabilization under standard electrical stimulation conditions, eugenol (5 mM) was added to the bath solution to determine whether eugenol was still able to increase the resting tension even after the preparation had been exposed to arsenite.

c) Effects of arsenite on saponin-permeabilized trabeculae

To determine whether arsenite was capable of developing strength or altering the sensitivity of contractile proteins to calcium, we developed a protocol with saponin-permeabilized fibers. After permeabilization, the preparations were transferred to solution A to obtain maximum strength, after stabilization under these experimental conditions, the trabeculae were transferred to solution R plus arsenite (7 mM) and left in these conditions for 20 minutes. After the incubation period, the preparation was transferred to solution A plus arsenite (7 mM), after stabilization, it was transferred to solution R, and then, after total relaxation, transferred to solution R plus eugenol (5 mM) and arsenite (7 mM), after stabilization, it was transferred to wash in solution R, and then transferred to solution R plus only eugenol (5 mM), to determine the control eugenol.

d) Effects of arsenite on trabeculae permeabilized by Triton X- 100

We designed a new experiment, this time using trabeculae permeabilized with Triton X-100, which would make it possible to determine the action of these agents directly on the contractile proteins. To carry out this protocol, we once again produced the experimental solution called SUCO (described in section 3.9), which is the product of the permeabilization of trabeculae by saponin over 80 minutes in R solution. Specifically, in this experimental protocol we used the following experimental solutions: solution A, solution A plus sodium arsenite (7 mM), SUCO plus eugenol (5 mM) and SUCO plus eugenol (5 mM) and sodium arsenite (7 mM). After being permeabilized by Triton X-100 for 10 minutes, the trabeculae were transferred to solution A to obtain the maximum force, after which they were transferred to solution A plus sodium arsenite (7 mM) to determine whether this compound alters the sensitivity of contractile proteins to Ca^{2} +. After preparation, the trabeculae were washed in R solution and then transferred to SUCO solution plus eugenol (5 mM). After the development of maximum resting tension, the trabeculae were transferred to SUCO solution plus eugenol (5 mM) and sodium arsenite (7 mM). A new set of trabeculae, permeabilized by saponin (20 minutes), was used to determine the control eugenol.

4.2 - Reduction in the concentration of intracellular ATP due to the action of eugenol

Studies carried out by Jeng et all (1994) show that eugenol is a biological agent capable of reducing the concentration of free intracellular ATP, so we carried out an experimental test to determine whether eugenol is capable of reducing the circulating concentration of ATP. The atria were dissected in normal ringer's solution, one of the atria was transferred to an eppendorf containing the same normal ringer's solution, while the other atrium was transferred to an eppendorf containing normal ringer's solution plus eugenol (5 mM). The preparations remained in these solutions for the following times: time zero (zero minutes), time 30 (30 minutes) and time 60 (60 minutes). After this time, the preparations were frozen in liquid nitrogen, macerated and weighed. The macerated tissue was then suspended in 1000 ml of perchlorate acid (0.4M) and homogenized, centrifuged for 5 min at 3000 rpm, with a constant temperature of 4° C. The supernatant was used for ATP measurements using the luciferase enzyme. To determine luciferase activity, 20 µl of each sample was pipetted in duplicate into a Falcon tube and read on a luminometer (GloMax - Promega, Fitchburg, WI, USA) using Promega Luciferase Assay Substrate. Protein concentrations in the samples were determined using a BioRad assay (Hercules, CA, USA). The data was normalized by the protein concentration.

5 - Drugs

All salts and substances used in this research were of analytical purity (eugenol purity = 99%) and purchased from Sigma Chemical Co. (St Louis, Mo; U.S.A.); Merck (Darmstadt, Germany) or Reagen (Rio de Janeiro, RJ, Brazil). Eugenol, procaine, 2,3-butanedione monoxide (BDM), ruthenium red, 1,4-dithiothreitol (DTT), colchicine and saponin were solubilized in deionized water and kept at a low temperature (stock solution), before being used in the experiments, they were kept at room temperature and constantly stirred on a magnetic stirrer. Sodium arsenite was solubilized in a solution containing 90% deionized water and 10% methanol and kept (stock solution) at -20°. Ryanodine was solubilized in 100% methanol and kept (stock solution) at -20°. The caffeine was weighed out before the experiments and added directly to the experimental solution in a glass container and stirred on a magnetic stirrer until it was completely solubilized, after which the volume was transferred to the experimental chamber. Triton X-100 was added directly to the experimental solution.

6 - Statistical Analysis

The results are presented as the mean + standard error of the mean of at least 6 observations (n). For statistical analysis we used Student's t-test and/or ANOVA (one-way or two-way). When ANOVA revealed a significant difference, Tukey's test was applied. Differences were considered statistically significant when the probability of

occurrence of the null hypothesis was < 0.05. Values of * $p<0.05$ and # $p<0.01$ were considered significant. Data analysis and the plotting of figures were carried out using GraphPad Prism System software (San Diego, CA, U.S.A.) and/or SigmaPlot and SigmaStat.

RESULTS

1 - Experiments with Indirect Electrical Stimulation: intact atrial muscle trabeculae

1.1 - Effects of Eugenol on the AEC process: dose-response curve

The typical records show that eugenol at different concentrations (1, 3, 5, 7 and 10 mM) was able to induce an increase in resting tension in a concentration-dependent manner. From this data, it was possible to determine the average value of shape production, i.e. LogEC50 of 4.781. The results show that eugenol reached its maximum resting tension at a concentration of 7 mM and that it stabilized at 10 mM.

	mN/mm²	N
1 mM	0,000±0,000	6
3 mM	2,963±0,735	6
5 mM	7,318±0,528	6
7 mM	10,862±1,812	6
10 mM	8,828±1,289	6

Table 2 - Effects of eugenol at different concentrations, dose response curve. Results are presented as mean ± SEM.

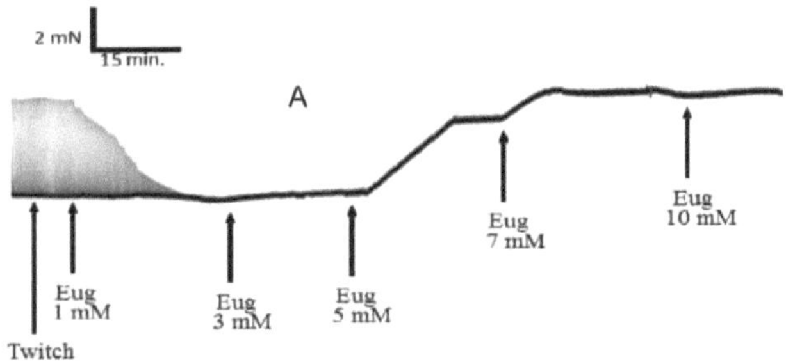

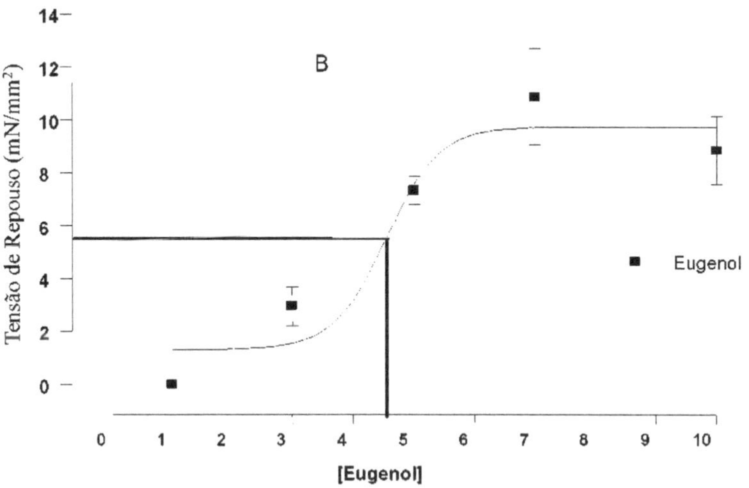

Figure 13 - (A): Typical recording of the dose-response effects of eugenol on the resting tension developed by rat atrial muscle trabeculae. (B): Effects of eugenol, at different concentrations, on the resting tension developed by atrial muscle trabeculae. Data are presented as mean EPM.

1.2 - Effects of Procaine on the action of Eugenol in normal ringer's solution

The results show that in normal Ringer's solution, procaine (10 mM) was able to abolish the contractions (tremor), but when eugenol (5 mM) was added to the bath solution, this compound was able to increase the resting tension, statistically the same as the control eugenol (5 mM).

	mN/mm²	N
Shake	7,989±1,035	6
Procaine	0,000±0,000	6
Eugenol + Procaine	7,698±0,933	6
Eugenol Control	7,318±0,529	6

Table 3 - Trabeculae treated with procaine. Experiments in normal ringer. Results are presented as mean ± SEM.

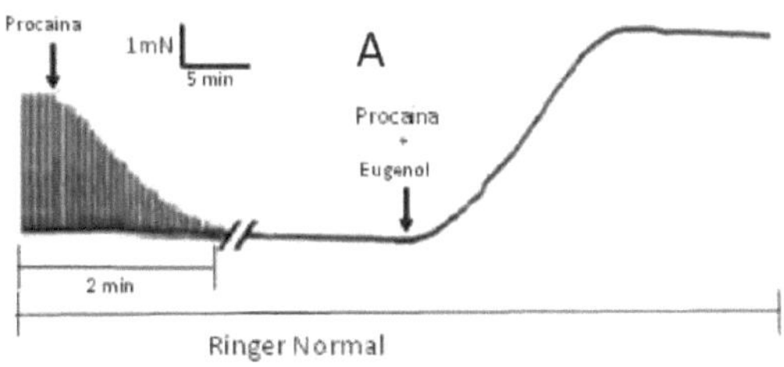

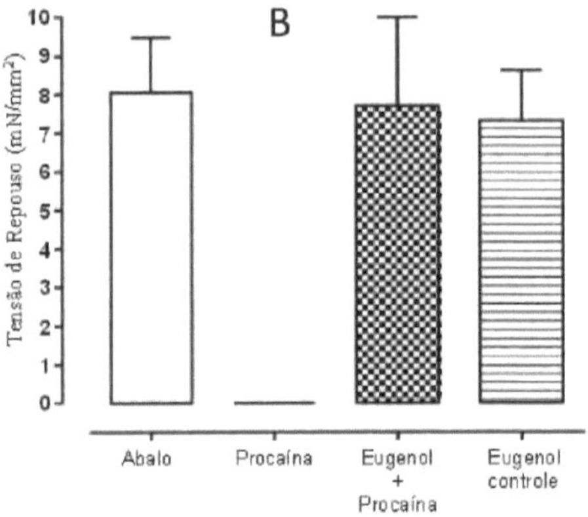

Figure 14 - Effects of eugenol after treatment with procaine on the resting tension developed by rat atrial muscle trabeculae in preparations kept in Normal Ringer's solution. A - typical record of the experiments; B - resting tension data. Data are presented as mean ± SEM.

1.3 - Effects of Procaine on the action of eugenol in zero-Ca ringer solution^{2+}

The results show that in zero-Ca ringer solution^{2+} eugenol (5 mM) was able to increase the resting tension even in the presence of procaine (10 mM), statistically the same as the control eugenol (5 mM).

	mN/mm^2	N
Shake	8,010±1,281	6
Procaine	0,000±0,000	6
Eugenol + Procaine	6,955±1,148	6
Eugenol Control	7,164±1,075	6

Table 4 - Trabeculae treated with procaine. Experiments in zero-Ca ringer^{2+} . Results are presented as mean ± EPM.

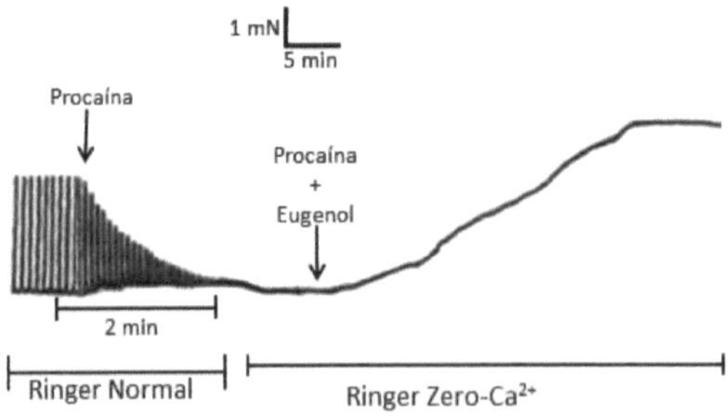

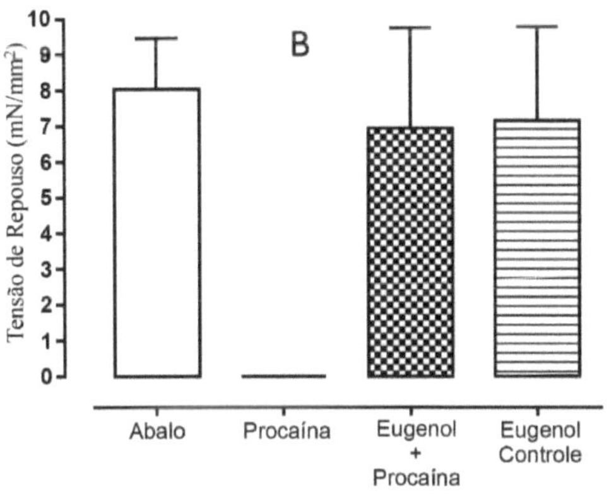

Figure 15 - Effects of eugenol after treatment with procaine on resting tension developed by rat atrial muscle trabeculae in preparations kept in zero-Ca2+ Ringer's solution. A - typical record of the experiments; B - resting tension data. Results are presented as mean ± SEM.

1.4 - Effects of Rianodine on the action of Eugenol in Normal Ringer's solution

The results show that ryanodine (100 µM) was able to abolish the contraction (jolt) induced by electrical stimulation. Even in the presence of ryanodine, eugenol (5 mM) was able to induce an increase in resting tension statistically equal to the control eugenol (5 mM).

	mN/mm²	N
Shake	8,080±1,109	6
Rianodina	0,000±0,000	6
Eugenol + Rianodine	7,329±1,436	6
Eugenol Control	7,318±0,529	6

Table 5 - Trabeculae treated with ryanodine. Experiments in normal ringer. Results are presented as mean ± EPM.

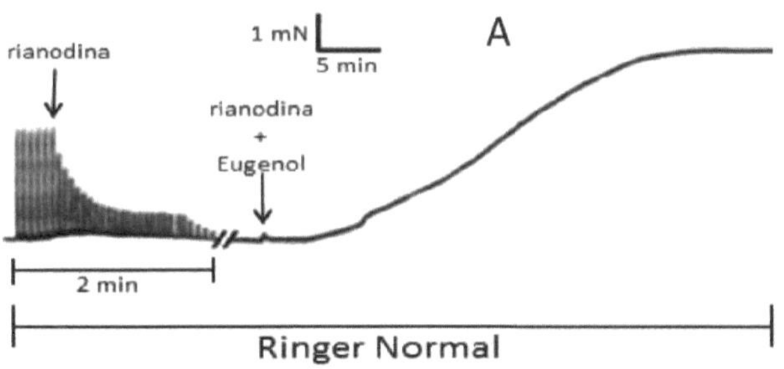

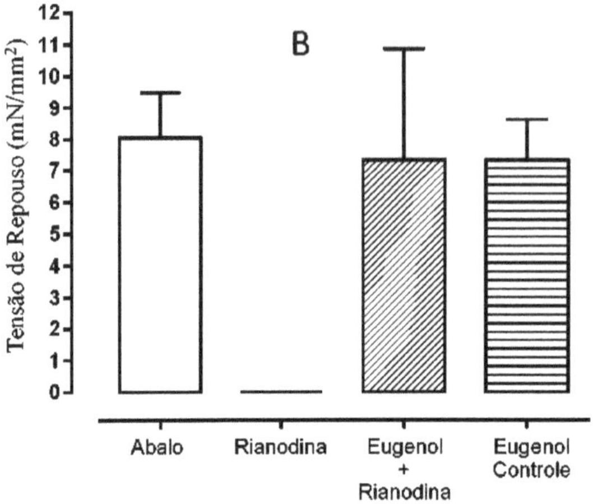

Figure 16 - Effects of eugenol after treatment with ryanodine on the resting tension developed by rat atrial muscle trabeculae in preparations kept in Normal Ringer's solution. A - typical record of the experiments; B - resting tension data. The results are presented as mean EPM.

1.5 - Effects of Rianodine on the action of Eugenol in zero-Ca Ringer's solution^{2+}

The results show that even in zero-Ca ringer solution^{2+} , eugenol (5 mM) was able to induce an increase in resting tension in the presence of ryanodine (100 µM), statistically the same as the control eugenol (5 mM).

	mN/mm²	N
Shake	7,899±1,075	6
Rianodina	0,000±0,000	6
Eugenol + Rianodine	7,420±0,911	6
Eugenol Control	7,164±1,075	6

Table 6 - Trabeculae treated with ryanodine. Experiments in zero-Ca ringer^{2+} . Results are presented as mean ± EPM.

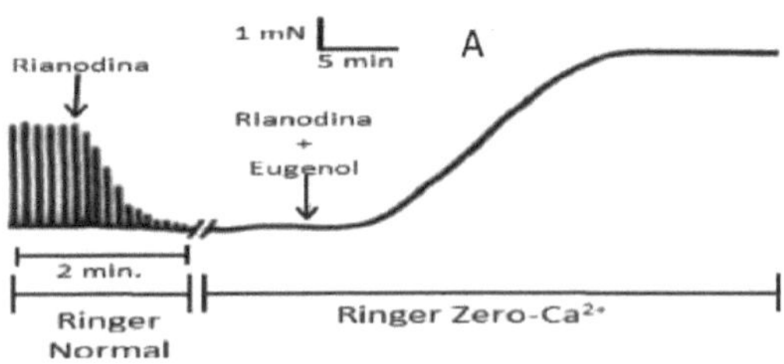

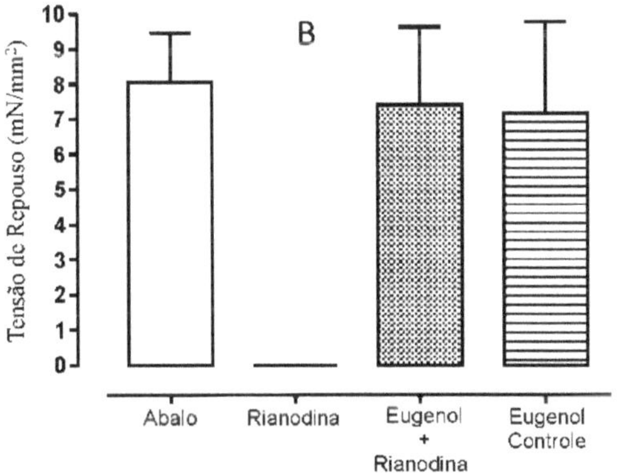

Figure 17 - Effects of eugenol after treatment with ryanodine on the resting tension developed by rat atrial muscle trabeculae in preparations kept in zero-Ca Ringer's solution^{2+} . A - typical record of the experiments; B - resting tension data. Results are presented as mean ± SEM.

1.6 - Effects of Rianodine and Procaine on the action of Eugenol in Normal Ringer's solution

The results show that even in the presence of two calcium release blockers, ryanodine (100 µM) and procaine (10 mM), eugenol (5 mM) was able to induce an increase in the resting tension of atrial trabeculae, equal to the tension developed with control eugenol (5 mM).

	mN/mm²	N
Shake	7,999±1,175	6
Rianodine + Procaine	0,000±0,000	6
Eugenol + Rianodine + Procaine	7,069±0,444	6
Eugenol Control	7,318±0,529	6

Table 7 - Trabeculae treated with procaine. Experiments in normal ringer. Results are presented as mean ± EPM.

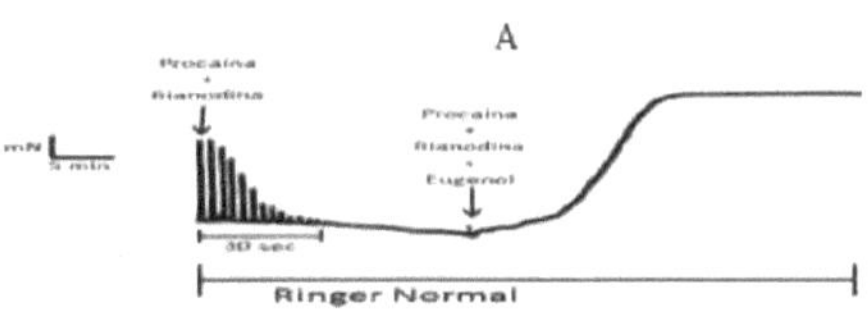

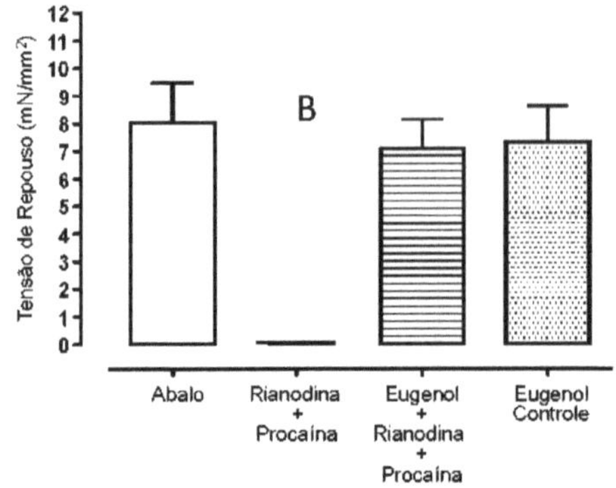

Figure 18 - Effects of eugenol after treatment with ryanodine and procaine on the resting tension developed by rat atrial muscle trabeculae in preparations kept in Normal Ringer's solution. A - typical record of the experiments; B - resting tension data. Results are presented as mean ± SEM.

1.7 - Effects of Rianodine and Procain on the action of Eugenol in zero-Ca Ringer's solution^{2+}

Even in zero-Ca ringer's solution^{2+} and in the presence of two calcium release blockers, ryanodine (100 µM) and procaine (10 mM), eugenol (5 mM) was able to induce an increase in the resting tension of atrial trabeculae, equal to the tension developed in control eugenol (5 mM).

	mN/mm^2	N
Shake	8,021±1,200	6
Rianodine + Procaine	0,000±0,000	6
Eugenol + Rianodine + Procaine	7,205±0,582	6
Eugenol Control	7,164±1,075	6

Table 8 - Trabeculae treated with procaine and ryanodine .Experiments in zero-Ca ringer^{2+} . Results are presented as mean ± SEM.

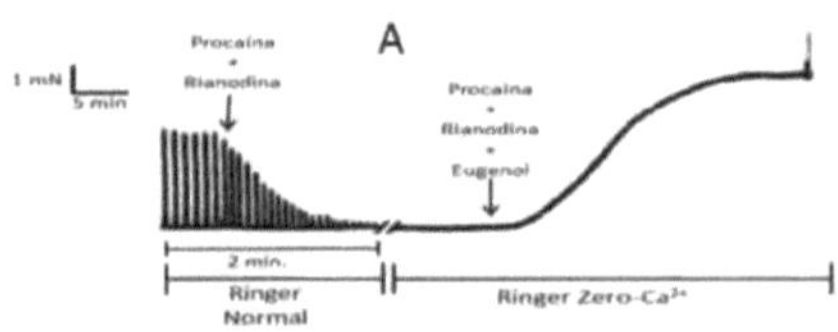

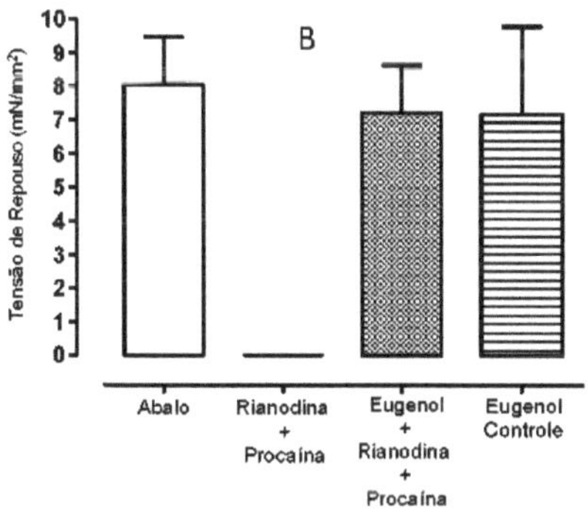

Figure 19 - Effects of eugenol after treatment with ryanodine and procaine on the resting tension developed by rat atrial muscle trabeculae in preparations kept in zero-Ca Ringer's solution^{2+} . A - typical record of the experiments; B - resting tension data. Results are presented as mean ± SEM.

1.8 - Effects of BDM on the action of Eugenol in Normal Ringer's solution

The results indicate that the increase in resting tension induced by eugenol (5 mM) is an active state of cross-bridge cyclization, since BDM (30 mM) was able to reduce resting tension. BDM-induced relaxation was achieved within 20 to 30 minutes of its addition to the solution.

	mN/mm^2	N
Procaine	0,000±0,000	6
Eugenol + Procaine	7,698±0,933	6
Eugenol + Procaine + BDM	0,283±0,209	6
Eugenol Control	7,318±0,529	6

Table 9 - Trabeculae treated with procaine and BDM. Experiments in zero-Ca ringer^{2+} . Results are presented as mean ± EPM.

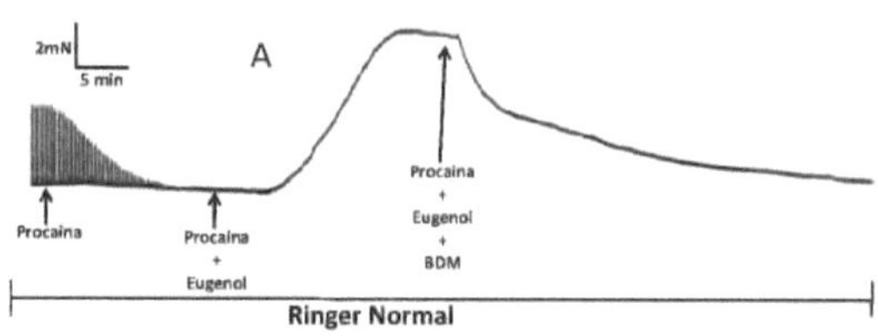

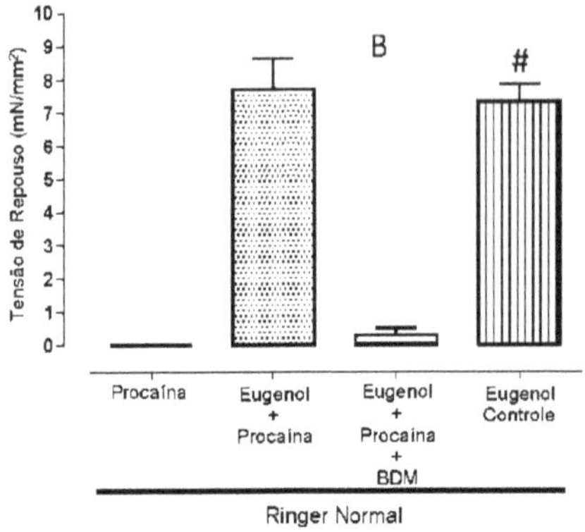

Figure 20 - Effects of eugenol after treatment with procaine and BDM on resting tension developed by rat atrial muscle trabeculae in preparations kept in Normal Ringer's solution. A - typical record of the experiments; B - resting tension data. Results are presented as mean ± SEM.

The results indicate that the increase in resting tension induced by eugenol (5 mM) is an active state of cross-bridge cyclization, since BDM (30 mM) was able to reduce resting tension, and that this increase is independent of the influx of extracellular Ca^{2+} . BDM-induced relaxation was achieved 20 to 30 minutes after its addition to the solution.

	mN/mm²	N
Procaine	0,000±0,000	6
Eugenol + Procaine	6,955±0,897	6
Eugenol + Procaine + BDM	0,055±0,095	6
Eugenol Control	7,164±1,075	6

Table 10 - Trabeculae treated with procaine and BDM. Experiments in zero-Ca ringer^{2+} . Results are presented as mean ± EPM.

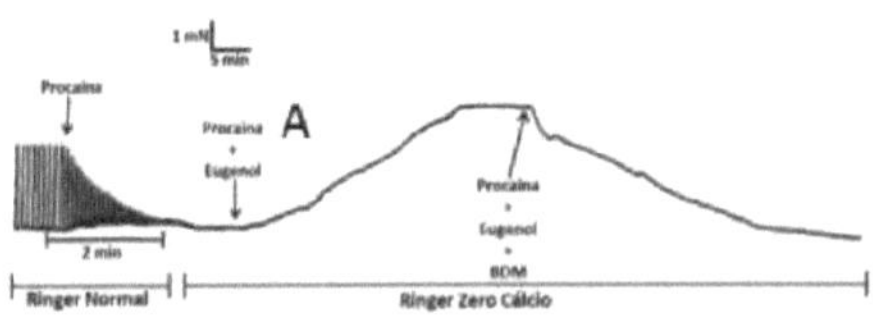

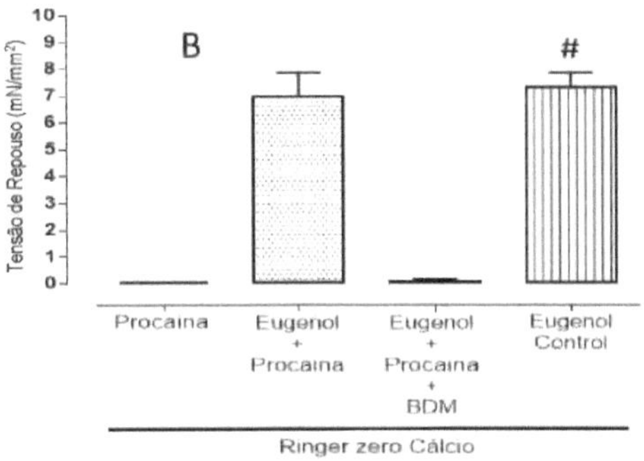

Figure 21 - Effects of eugenol after treatment with procaine and BDM on resting tension developed by rat atrial muscle trabeculae in preparations kept in zero-Ca Ringer's solution^{2+} . A - typical record of the experiments; B - resting tension data. Results are presented as mean ± SEM.

1.10 - Effects of Eugenol after depletion of calcium stored in the sarcoplasmic reticulum by caffeine in zero-Ca Ringer^{2+} and zero-Ca Ringer /Zero-Na^{2++}

Confirming the results of the previous protocols, even after calcium depletion of the SR induced by caffeine (30 mM), eugenol (5 mM) was able to induce an increase in resting tension. The activity of the Na^+ / Ca^{2+} exchangers was controlled depending on the experimental solution (zero-Ca ringer^{2+}/Na^+ and zero-Ca ringer^{2+}), making it possible to identify that the action of eugenol (5 mM) is not via this route. In short, the pathway by which eugenol acts is independent of extra or intracellular Ca^{2+} .

	mN/mm^2	N
Ringer zero-Ca^{2+}/Na$^+$	0,000±0,000	8
Caffeine (1)	3,641±0,976	8
Caffeine (2)	0,482±0,252	8
Caffeine (3)	0,135±0,135	8
Eugenol+Caffeine	7,214±2,128	8
Eugenol Control	7,318±0,529	8

Table 11 - Experiments to deplete Ca^{2+} in RS by the action of caffeine. Results are presented as mean ± SEM.

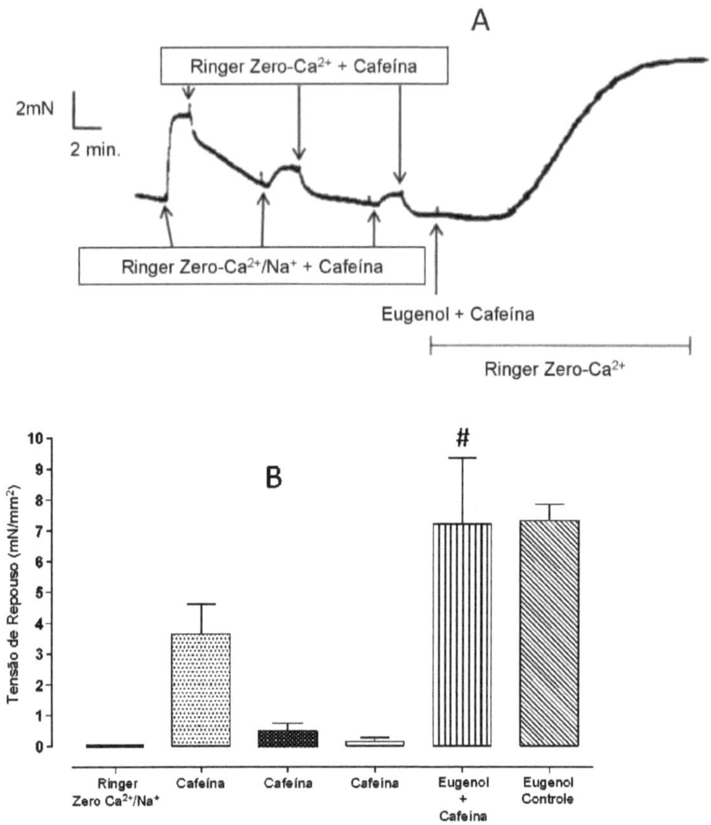

Figure 22 - Effects of eugenol after depletion of calcium stored in the SR by caffeine (30 mM) on the resting tension developed by rat atrial muscle trabeculae in preparations kept in zero-Ca Ringer's solution^{2+} . A - typical record of the experiments; B - resting tension data. Results are presented as mean ± SEM; # $p<0.01$; ANOVA (one-way) followed by Tukey's test, n=8; Eugenol+Caffeine *vs* Caffeine.

2 - Experiments with rat atrial muscle trabeculae permeabilized by saponin and/or Triton X-100

2.1 - Effects of eugenol on resting tension in atrial trabeculae permeabilized by Saponin and Triton X-100

The results presented in solution A (pCa4.0) indicate the maximum force of the preparations. The force produced in caffeine (30 mM) represents the activation of the contractile machinery by Ca^{2+} released from the SR. Eugenol (5 mM) was able to induce an increase in resting tension in R solution (high EGTA), but when the trabeculae were permeabilized with Triton X-100 the tension induced by eugenol was reversed.

	mN/mm²	N
pCa 4.0	13,120±1,355	8
Caffeine	4,140±0,678	8
Eugenol	6,697±1,147	8
Eugenol+Triton X-100	-0,710±0,298	8

Table 12 - Experiments with atrial trabeculae permeabilized by saponin and Triton X-100. Results are presented as mean ± SEM.

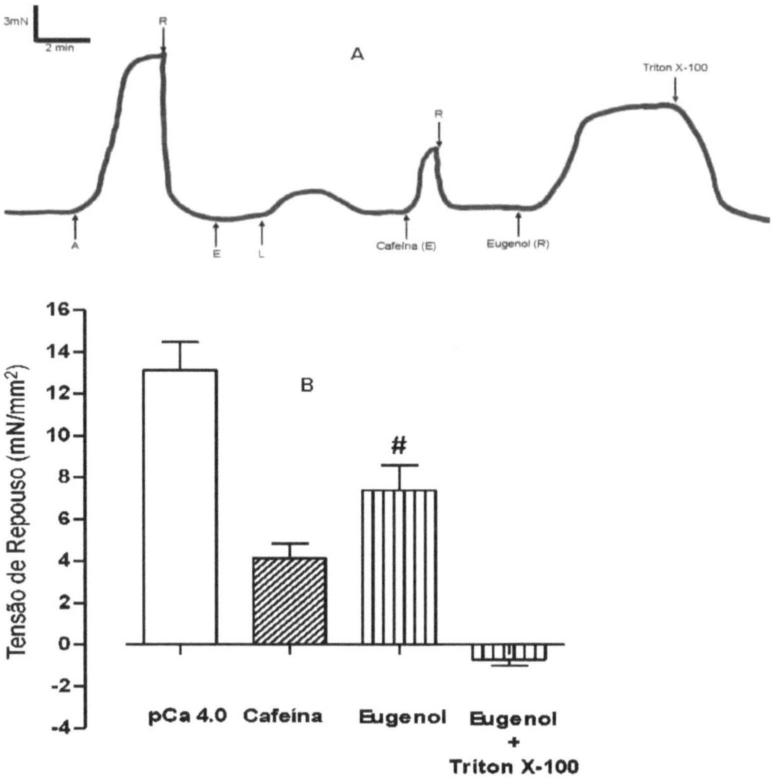

Figure 23 - Effects of eugenol on preparations of atrial trabeculae permeabilized with saponin (30 mg/ml) for 20 minutes. A - typical recording of the experiments; B - resting tension data. Results are presented as mean ± SEM; # p<0.01 ; ANOVA and Tukey; n = 8; Eugenol *vs* Caffeine; Eugenol *vs* Eugenol+Triton X-100.

2.2 - Effects of ruthenium red (RR) on the action of eugenol on saponin-permeabilized trabeculae

Eugenol in the presence of ruthenium red was able to induce an increase in resting tension that was statistically the same as the control eugenol, reinforcing the results obtained previously, indicating that the pathway through which eugenol acts is independent of RyRs and through a Ca-independent pathway .$^{2+}$

	mN/mm²	N
pCa 4.0	14,515±1,589	8
Eugenol + Ruthenium Red	7,824±1,874	8
Eugenol Control	7,380±1,205	8

Table 13 - Experiments with atrial trabeculae permeabilized by saponin and treated with ruthenium red. Results are presented as mean ± SEM.

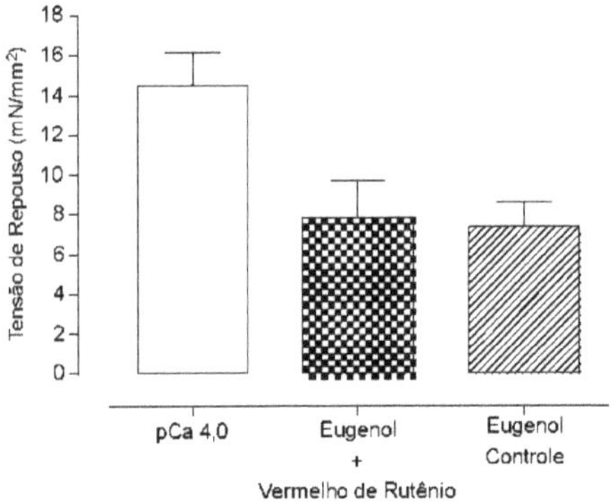

Figure 24 - Effects of eugenol (5 mM) on preparations of permeabilized atrial trabeculae treated with ruthenium red (10 µM). Results are presented as mean ± SEM.

2.3 - Effect of BDM on the action of eugenol on trabeculae permeabilized by Saponin

The experiments carried out with BDM (10 mM) and BDM (30 mM) showed that this compound induced relaxation of the preparations in the presence of eugenol (5 mM), although the higher the concentration of BDM, the greater the relaxation. With a statistical difference when compared to the control eugenol (5 mM)

	mN/mm²	N
Eugenol	7,222±0,781	8
Eugenol + BDM (10mM)	2,635±0,153	8
Eugenol + BDM (30 mM)	0,162±0,192	8
Eugenol Control	7,380±1,205	8

Table 14 - Experiments with atrial trabeculae permeabilized by saponin and treated with ruthenium red. Results are presented as mean ± SEM.

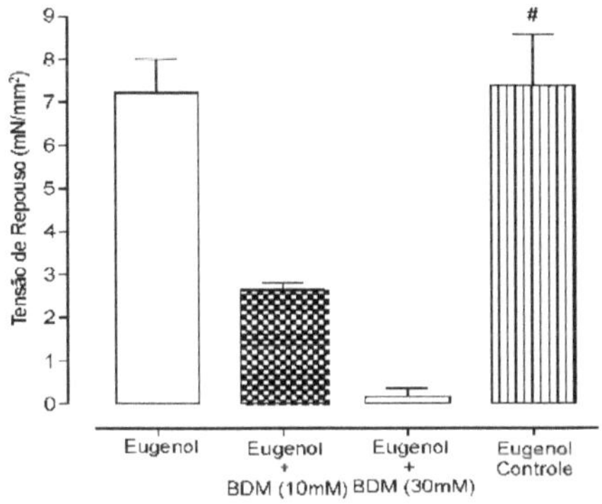

Figure 25 - Effects of eugenol (5 mM) on preparations of atrial trabeculae in the presence of BDM (10 mM) and (30 mM). The results are presented as mean EPM; # p<0.01 ; Student's t-test; n = 8; Eugenol+BDM 10 mM vs Eugenol Control; Eugenol+BDM 30 mM vs Eugenol Control.

2.4 - Relationship between strength and calcium concentration in the presence and absence of eugenol

The results show that eugenol (5 mM) is unable to alter the sensitivity of contractile proteins to Ca^2 in preparations permeabilized by Triton X-100.

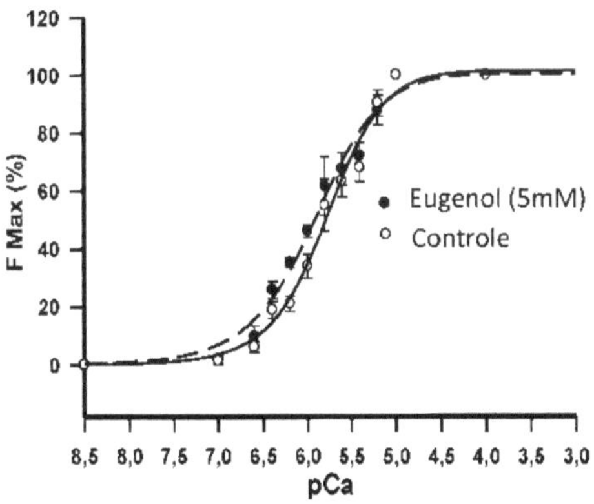

Figure 26: Relationship between strength at different concentrations of Ca^{2+} , in the absence and presence of Eugenol in trabeculae permeabilized with Triton X-100. The strength data was normalized to percentages and presented as means and standard deviations. The strength values in the absence and presence of eugenol were 1.37±0.03; 1.60±0.2 µM and 1.18±0.02; 1.24±0.17 µM (n=6).

2.5 - Effects of eugenol on resting tension of atrial trabeculae before and after permeabilization with Triton X-100

The results show that the Ca ions^2 + liberated by the presence of caffeine (30 mM) were completely chelated by EGTA (5 mM), demonstrating the effective action of this compound. Even in R solution, eugenol (5 mM) was able to induce an increase in resting tension in saponin-permeabilized trabeculae. However, when the trabeculae were permeabilized with Triton X-100, the response to eugenol (5 mM) disappeared, and the same was not observed in solution A, indicating that Triton X-100 does not alter the cyclization capacity of the cross-bridges.

	mN/mm²	N
pCa 4.0	13,819±1,640	8
Caffeine	0,001±0,000	8
Eugenol	6,710±1,045	8
Triton X-100	0,000±0,000	8
Eugenol (after Triton)	0,094±0,079	8
pCa 4.0 (post Triton)	11,477±1,812	8

Table 15 - Experiments with atrial trabeculae permeabilized by saponin and Triton X-100, submitted to caffeine in solution R, eugenol before and after Triton X-100 and solution A before and after Triton X-100. The results are presented as mean ± SEM.

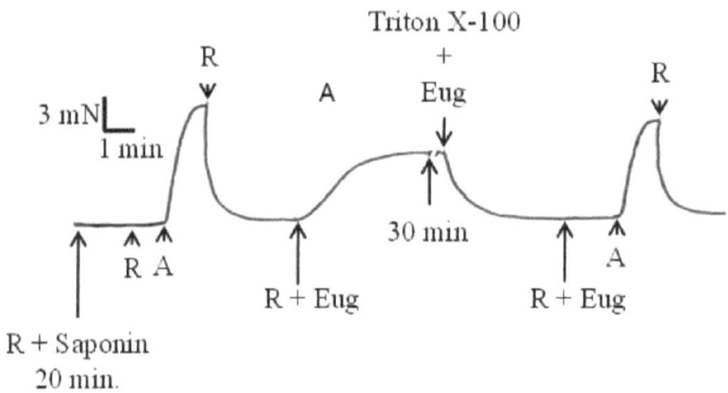

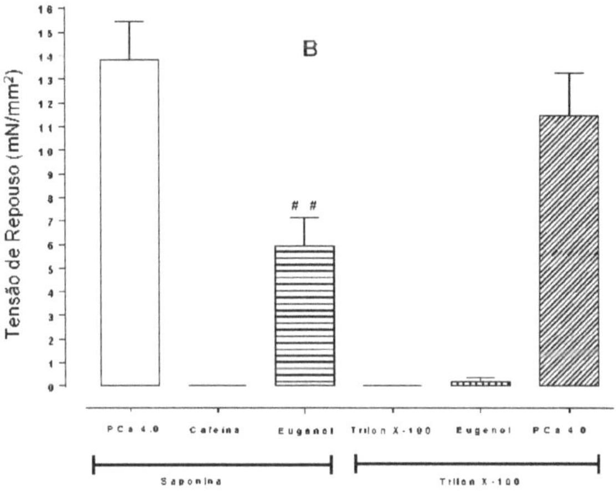

Figure 27 - Effects of eugenol on preparations of atrial trabeculae permeabilized with saponin (30 mg/ml) for 20 minutes and Triton X-100 (1% - vol/vol) for 10 minutes. A - typical recording of the experiments; B - resting tension data. The results are presented as mean EPM; # $p<0.01$; t-Student test; n = 8; Eugenol *vs* Eugenol post Triton X-100; Eugenol vs Caffeine.

2.6 - Effects of saponin exposure time on calcium- and caffeine-induced force production capacity and Eugenol-induced increase in resting tension

The results showed that permeabilization is necessary to obtain the maximum form. The response to caffeine (30 mM) occurred without membrane permeabilization (time zero), but the magnitude of the response was considerably greater after permeabilization. In relation to eugenol (5 mM), we observed a decrease in the response, i.e. in the ability of this oil to induce an increase in resting tension. No response was observed at 80 minutes.

Time	pCa 4.0	Caffeine	Eugenol	N
Zero (intact)	0,000±0,000	1,714±0,578	12,805±0,957	6
20 minutes	15,330±1,281	5,177±0,217	8,055±1,115	6
40 minutes	14,557±1,581	4,462±0,434	3,850±0,681	6
60 minutes	15,930±4,162	4,273±0,875	1,565±0,389	6
80 minutes	16,477±3,165	4,732±0,292	0,085±0,038	6

Table 16 - Trabeculae permeabilized at different times (zero, 20, 40, 60 and 80 minutes) by saponin. Relationship between permeabilization time and force and/or resting tension developed. Results are presented as mean ± SEM.

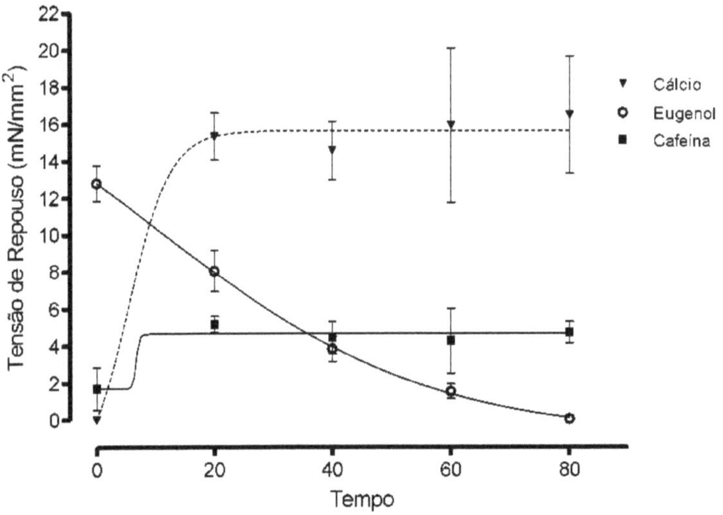

Figure 28 - Effects of Eugenol (5 mM) on trabeculae subjected to different exposure times to saponin (30µg/ml). Mechanical force production data in Solution A, Solution E plus Caffeine (30 mM) and Solution R plus Eugenol (5 mM). Results are presented as mean ± SEM.

2.7 - Effects of 1,4-Dithiothreitol (DTT) on the action of Eugenol

In the experiments carried out with DTT (100 µM), we observed that the increase in resting tension induced by Eugenol (mM), in the presence of this compound and at a high concentration of EGTA, was greater than the control eugenol (5 mM).

	mN/mm^2	N
pCa 4.0	13,120±1,355	8
Caffeine	4,140±0,678	8
DTT	0,000±0,000	8
Eugenol + DTT	13,595±1,854	8
Eugenol Control	7,380±1,205	8

Table 17 - Experiments with atrial trabeculae permeabilized by saponin and treated with DTT. Results are presented as mean ± SEM.

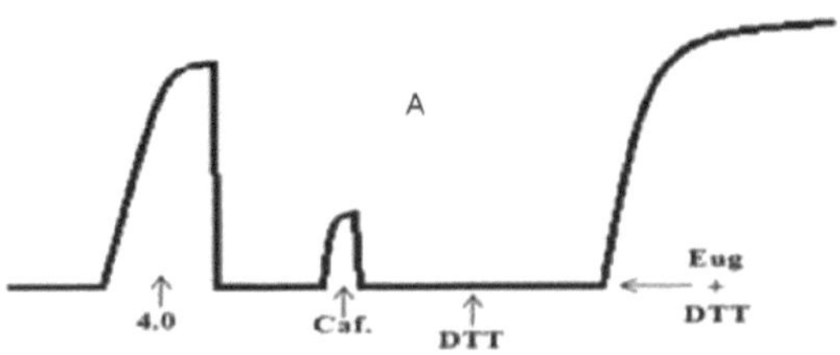

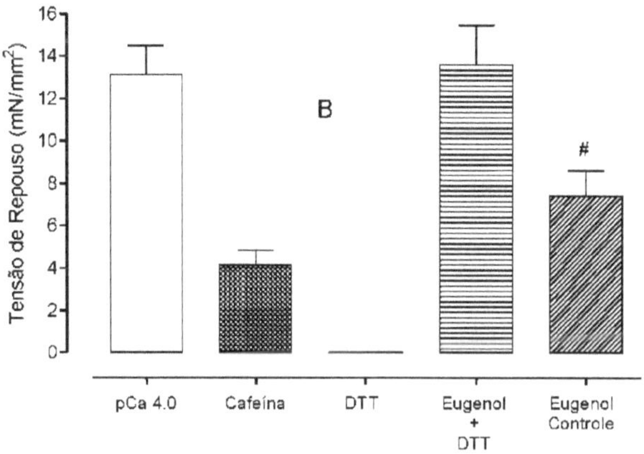

Figure 29 - Effect of Eugenol (5 mM) on trabeculae permeabilized with saponin (30µg/ml) and incubated with DTT (100 µM) (Dithiothreitol) for 20 minutes. A - typical recording of the experiments; B - resting tension data. The results are presented as mean EPM; # $p<0.01$; t-Student test; n = 8; Eugenol vs Eugenol+DTT.

2.8 - Effects of Colchicine (colch) on the action of Eugenol

The results show that after treatment with colchicine (30 µM), eugenol (5 mM) was able to induce an increase in resting tension to values higher than those observed with control eugenol (5 mM), indicating that the action of this compound somehow potentiated the action of eugenol.

	mN/mm^2	**N**
pCa 4.0	13,428±0,528	7
Caffeine	4,678±0,853	7
Colchicine	0,000±0,000	7
Eugenol + Colchicine	10,413±0,324	7
Eugenol Control	7,554±0,878	7

Table 18 - Experiments with atrial trabeculae permeabilized by saponin and treated with Colchicine. Results are presented as mean ± SEM.

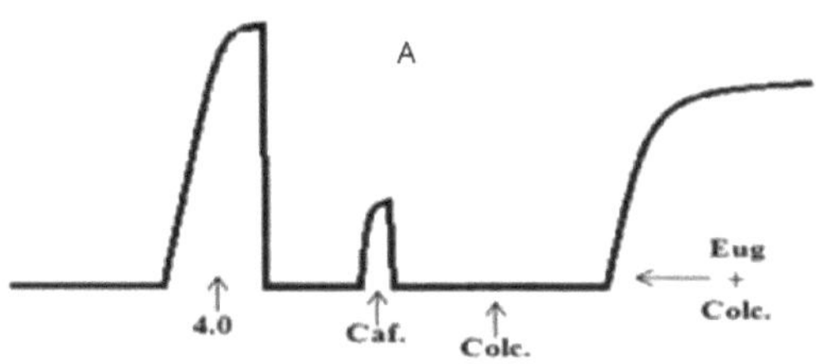

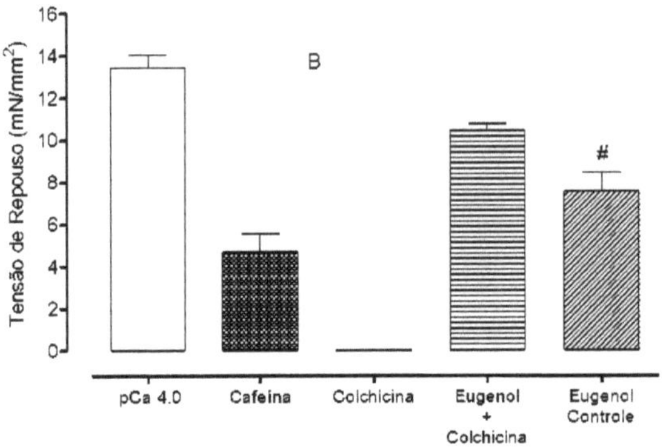

Figure 30 - Effect of Eugenol on trabeculae permeabilized with saponin (30µg/ml) and treated with Colchicine (30µM) for 20 minutes. A - typical recording of the experiments; B - resting tension data. The results are presented as mean EPM; * $p<0.05$; t-Student test; n = 7; Eugenol vs Eugenol+Colchicine.

2.9 - Experiments with the product of the permeabilization of trabeculae and its effect on Eugenol

These data indicate that there has been "reincorporation" of the proteins, that only the protein SUCO is able to increase the resting tension, but in the presence of Eugenol (5 mM) the resting tension is statistically equal to the control eugenol (5 mM). Only the SUCO+COLCH solution was able to induce an increase in resting tension equal to the control eugenol (5 mM), and when we added eugenol (5 mM) to this solution we observed an increase in resting tension greater than that shown by the control eugenol (5 mM).

	mN/mm²	N
pCa 4.0	13,428±0,582	8
Caffeine	0,000±0,000	8
Eugenol	0,000±0,000	8
Juice (saponin)	1,360±0,152	8
Juice (saponin) + eugenol	5,702±0,493	8
Juice (sap and colch)	5,794±0,784	8
Juice (sap and colch) + eugenol	8,786±0,673	8
Eugenol control	6,835±0,526	8

Table 19 - Reincorporation experiments of the product of permeabilization for 80 minutes in trabeculae permeabilized by Triton X-100. The results are presented as mean ± EPM.

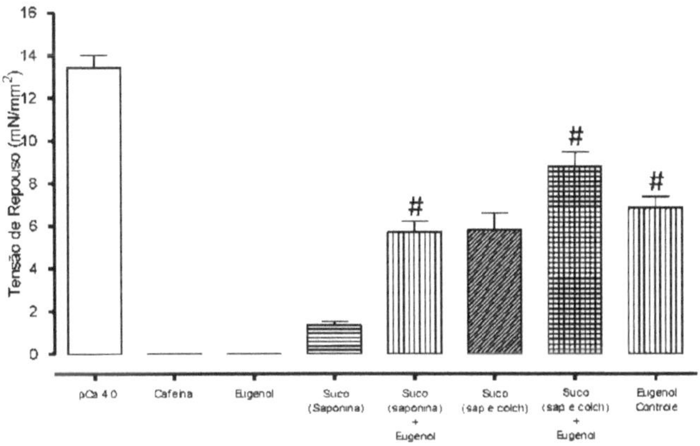

Figure 31 - Effects of Eugenol (5 mM) on the resting tension developed by atrial trabeculae permeabilized with Triton X-100 (1% vol/vol) in solution containing the product of saponin permeabilization (80 minutes) and solution containing the product of saponin permeabilization (80 minutes) and colchicine (30 µM). The results are presented as mean ± SEM; SUCO vs SUCO + Eugenol; SUCO (Colch) vs SUCO (Colch) + Eugenol; SUCO + Eugenol vs SUCO (Colch) + Eugenol (*p<0.05; # p<0.01; N = 8).

3 - Experiments to identify the possible routes of action of Eugenol

3.1 - Possible involvement of the p38 MAPk pathway

3.1.1 - Dose response curve to sodium arsenite

Our results show that arsenite (1, 3, 5, 7 and 10 mM) was able to reverse the resting tension induced by eugenol (5 mM) in a concentration-dependent manner. Based on these data, it was possible to determine the concentration of this compound to be used in our experiments (EC50).

[Arsenite]	mN/mm²	N
1 mM	7,609±1,312	8
3 mM	9,421±0,916	8
5 mM	7,788±1,198	8
7 mM	2,885±0,718	8
10 mM	0,000±0,000	8

Table 20 - Dose response curve to arsenite in the presence of eugenol (5 mM) in intact trabeculae. The results are presented as mean ± EPM.

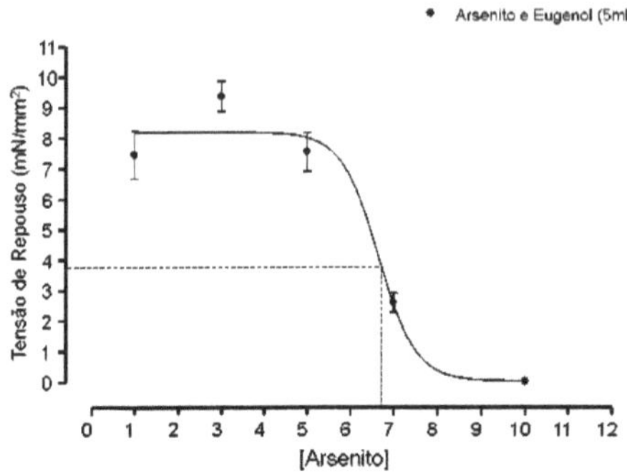

Figure 32 - Effects of arsenite in the presence of eugenol (5 mM), dose response curve. The experiments were carried out on intact trabeculae and normal Ringer's solution (IC50=6.78 mM). The results are presented as mean ± SEM, n=8.

3.1.2 - Arsenite experiments: recovery after arsenite action

The results show that sodium arsenite (7 mM) was able to abolish the twitch induced by indirect electrical stimulation. After washing the preparation, the trabeculae were able to respond to electrical stimulation, producing less force than that observed before incubation with sodium arsenite. Eugenol (5 mM) was able to induce an increase in resting tension after washing the preparation.

	mN/mm²	**N**
Twitch	4,487±0,399	8
Arsenite (7 mM)	0,000±0,000	8
Recovery	1,982±0,575	8
Eugenol (5 mM)	5,198±0,321	8

Table 21- Results of the experiments to determine whether the toxicity of sodium arsenite could be determining the response to eugenol. The results are presented as mean ± EPM.

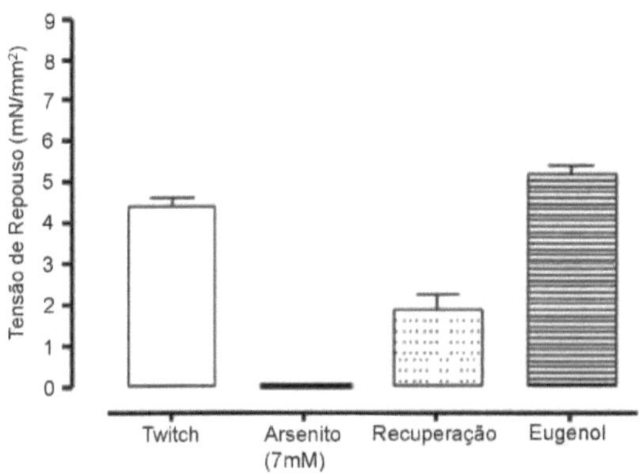

Figure 33 - Effects of Eugenol in the presence and absence of Arsenite. The trabeculae were electrically stimulated and Arsenite (7 mM) was added to the solution (according to IC50). After 20 minutes, the preparation was washed and stimulated again. Eugenol (5 mM) was then added to the normal ringer solution. The experiments were carried out on intact trabeculae. Results are presented as mean ± SEM. n=8.

3.1.3 - Arsenite experiments: fibers permeabilized with saponin

Our results showed that, in preparations permeabilized with saponin, sodium arsenite (7 mM) was unable to alter the cyclization of cross-bridges in the presence of Ca^{2+} (solution A). In permeabilized trabeculae, the ability of eugenol (5 mM) to increase the resting tension was reduced in the presence of sodium arsenite (7 mM), a result that was statistically lower than that of the control eugenol (5 mM).

	mN/mM^2	N
pCa 4.0	12,185±1,890	8
Arsenite (7 mM)	2,615±0,420	8
pCa 4.0 + arsenite (7 mM)	12,301±1,509	8
Eugenol (5 mM) + arsenite (7 mM)	2,683±0,321	8
Eugenol control	6,912±1,010	

Table 22- Experiments with saponin-permeabilized atrial trabeculae. Incubation with sodium arsenite (7 mM). Results are presented as mean ± SEM.

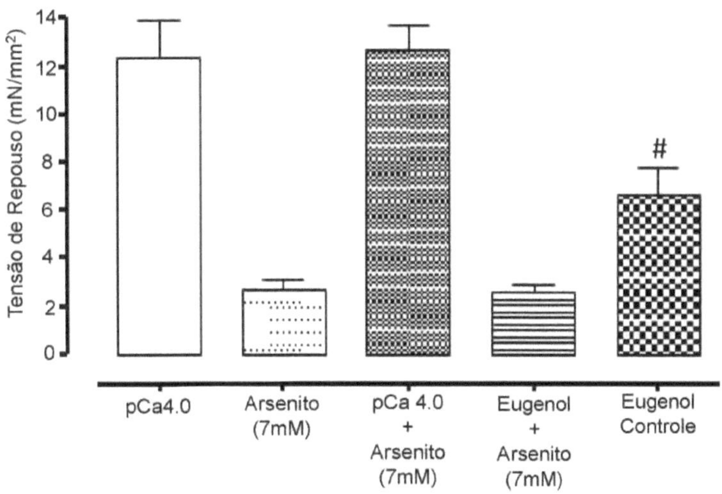

Figure 34 - Effects of Eugenol in the presence and absence of Arsenite. The experiments were carried out with trabeculae permeabilized by saponin (30µg/ml) for 20 minutes; incubated in arsenite (7 mM) for 30 minutes. The results are presented as mean EPM; # p<0.01; Student's t-test, n=8; pCa 4.0 *vs* pCa4.0+Arsenite; Eugenol+Arsenite *vs* Eugenol Control.

3.1.4 - Effects of arsenite on trabeculae permeabilized by Triton X-100

In preparations permeabilized with T riton X-100, arsenite was unable to alter the cyclization of cross-bridges in the presence of Ca^{2+} (solution A). In SUCO solution, eugenol (5 mM) was able to induce an increase in resting tension, but when we added sodium arsenite (7 mM) to the solution, it was able to reverse the resting tension induced by eugenol. The results of eugenol in SUCO solution plus sodium arsenite are statistically lower than the control eugenol (5 mM).

	mN/mm^2	N
pCa 4.0	12,800±1,832	8
pCa 4.0 + arsenite (7 mM)	14,132±1,799	8
Eugenol (5 mM) + Juice	7,045±1,909	8
Eugenol (5 mM) + SUCO + arsenite (7 mM)	1,125±0,281	8
Eugenol control	6,987±1,032	

Table 23- Experiments with atrial trabeculae permeabilized by Triton X-100, using the solution with the product of the permeabilization of trabeculae for 80 minutes in saponin. The experiments were carried out in the presence and absence of sodium arsenite (7 mM). The results are presented as mean ± SEM.

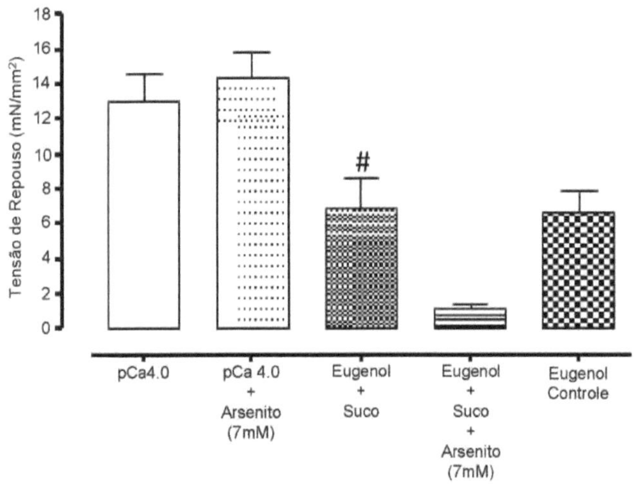

Figure 35 - Effects of Eugenol in the presence and absence of Arsenite. The trabeculae were permeabilized with Triton X-100 (1% - vol/vol) for 10 minutes. The trabeculae were exposed to pCa 4.0 in the absence and presence of Arsenite and exposed to Eugenol + SUCO in the presence and absence of Arsenite. The results are presented as mean EPM; # $p<0.01$; Student's t-test, n = 8; Eugenol+Suco *vs* Eugenol+Suco+Arsenite.

3.2 - Reduction in the concentration of intracellular ATP due to the action of eugenol

The results showed that eugenol is able to reduce the concentration of ATP in a time-dependent manner.

	µMol ATP/g	N
C 0	0,882±0,079	5
T 0	0,798±0,143	5

C 30	0,844±0,114	5
T 30	0,524±0,097	5
C 60	0,849±0,026	5
T 60	0,346±0,112	5

Table 24 - ATP dosage, values expressed in µM of ATP per gram of protein. ATP concentration was measured in the absence (C) and presence (T) of eugenol (5mM) at 0, 30 and 60 minutes. The results are presented as mean ± SEM.

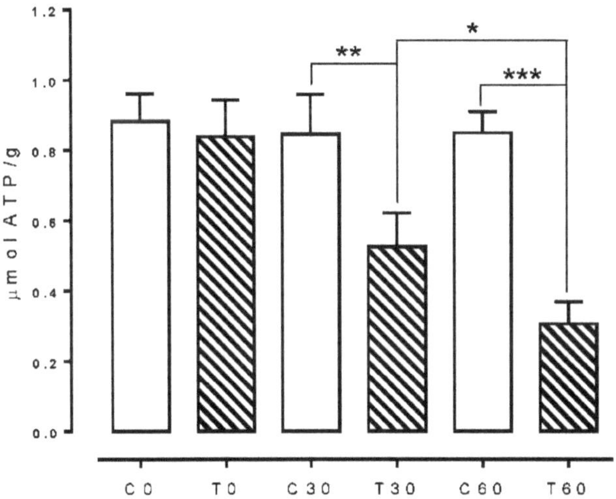

Figure 36 - The effect of eugenol on ATP concentration in atrial muscle. ATP concentration was measured in the absence (C) and presence (T) of eugenol at 0, 30 and 60 minutes. The values are expressed as µmol of ATP per g of protein. Symbols represent differences between groups ($p < 0.05$): * E 30 vs E 60; ** C30 vs E30; *** C60 vs E 60.

DISCUSSION

The results obtained in this study show that, in atrial muscle trabeculae submitted to electrical stimulation, eugenol was able to completely inhibit the isometric contraction developed. Furthermore, at the concentrations used, eugenol was able to induce an increase in resting tension even in the presence of RyR blockers. In the experiments with saponin-permeabilized atrial muscle trabeculae, eugenol was able to induce an increase in resting tension, under experimental conditions that lead us to believe that it is a calcium-independent pathway, but when the preparations were treated with Triton X-100, the response to eugenol disappeared, indicating that this oil does not act directly on contractile proteins and that it possibly depends on a soluble cytosolic agent.

The events that occur between the action potential in the muscle fiber and its contraction are defined as the Excitation-Contraction Coupling (ECC) process (Bers, 2002). Among the ions involved in the complex working mechanism of the heart, calcium can perhaps be considered the most important. It is crucial for the mechanisms that enable the heart's cardiac chambers to contract and relax, a process called excitation-contraction coupling (Bers, 2002).

It is a fact that calcium plays a fundamental role in triggering cardiac contraction, as first suggested by Ringer in 1883 (Ringer, 1883). A descriptive model has been formulated to explain the functional properties of the cardiac cell which involves all the events that begin with the action potential and result in an increase in Ca^{2+} and the activation of contractile proteins (Schouten et al., 1989). In atrial myocytes, the release of calcium from the SR during the AEC process occurs primarily through the RyRs (Kockskamper et al., 2001; Sheehan and Blatter, 2003).

Besides being the main constituent of clove oil, eugenol is also a constituent of the essential oils or extracts of many other plants, such as cinnamon and basil (Thompson et al., 1989; Wrigh et al., 1995). Over the years, eugenol has been used as a valuable tool in an attempt to better elucidate the events related to the AEC process. Work carried out on rat papillary muscles showed that eugenol, at low concentrations (0.01 - 0.5 mM), has a negative inotropic action, acting as a type-L Ca^{2+} channel blocker, but is unable to affect the activity of the myosinic Ca^{2+} -ATPase (Damiani et al., 2004).

In another study, the effects of eugenol at high concentrations (1 - 10 mM) on the SR as well as on the contractile apparatus of chemically dismembered rat skeletal muscle fibers were investigated. The results showed that eugenol (5 mM) was able to induce contractions possibly through the release of Ca^{2+} from the SR. Furthermore, the authors concluded that the muscle contraction induced by eugenol involved at least two mechanisms for the release of Ca^{2+} by the SR: a mechanism related to the activation of RyRs and another mechanism through a heparin-sensitive pathway or, more precisely, an IP3Rs pathway (Lofrano-Alves et al., 2005).

The experiments with intact atrial muscle trabeculae and under paced stimulation showed that eugenol at different concentrations (1 -10 mM) was able to abolish muscle jerking and, in a concentration-dependent manner, was able to induce an increase in the resting tension line. This event therefore characterizes the ability of eugenol, at the concentrations used, to induce an increase in resting tension in atrial muscle. These results are similar to those found in preparations with rat skeletal muscle (Lofrano-Alves, et al., 2005).

In the heart muscle, the strength of contraction is dependent on the extracellular concentration of Ca^{2+} , $[Ca^{2+}]_o$. It is currently accepted that the calcium required for contraction comes from two sources: 1) the influx of extracellular calcium via the slow Ca^{2+} channels (DHPRs) and, 2) the Ca^{2+} store in the sarcoplasmic reticulum (RyRs, mainly). Calcium from the extracellular environment passes into the cell and triggers the release of Ca^{2+} from the sarcoplasmic reticulum. It is estimated that around 30% of the Ca^{2+} required to raise the intracellular concentration of Ca^{2+} , $[Ca^{2+}]_i$, necessary for the development of maximal strength, is derived from the flow of Ca^{2+} through the sarcolemma and the remaining 70% represents the activating Ca^{2+} derived from the sarcoplasmic reticulum (Sperelakis *et al.,* 1996). A lot of information can be gleaned from the records of experiments in which we have used a drug that blocks calcium-activating channels in the SR. It has been described that inhibition of Ca^{2+} influx by eugenol (Damiani et. al., 2004) or by Ni^{2+} results in a high concentration of Ca^{2+} released from the sarcoplasmic reticulum (Komai & Rusy, 1993). This reasoning is exactly the opposite of the notion that inhibition of Ca^{2+} influx would invariably result in a decrease in the concentration of Ca^{2+} in the SR.

Our hypothesis, based on previous results, is that the increase in resting tension induced by eugenol is the result of a Ca^{2+} independent pathway. One of the main effects of procaine on muscle cells is its ability to inhibit the release of Ca^{2+} by the SR (Endo, 1977; Stephenson & Wendt, 1986). Thus, we used procaine to observe the effects of eugenol on calcium release by the SR (Zahradnikovà & Palade, 1993). Our results show that procaine was able to completely block the isometric contraction (jolt) developed by atrial muscle trabeculae under indirect stimulation; however, when eugenol was added to the solution, it was able to induce an increase in resting tension. These results show that procaine effectively produced a blocking effect on the RyRs of the sarcoplasmic reticulum, but when eugenol was added to the trabecula, it was able to increase resting tension. The results therefore suggest that eugenol is acting through a different pathway to RyRs.

The AEC process involves L-type voltage-dependent calcium channels (DHPRs) which allow calcium to enter and whose activity results in the release of calcium via RyRs (Xu et al., 1994) or another calcium-releasing channel from the SR (Moore et al., 2004). Thus, the use of a solution without calcium (zero-Ca ringer's solution^{2} +) gave us the certainty that extracellular calcium was not effectively contributing to the calcium-induced calcium release process. Furthermore, the addition of procaine to the solution would not allow calcium to be released via the RyRs of the SR. However, when eugenol was added to the preparation under these experimental conditions, it was once again able to induce an increase in resting tension. These results show that despite the fact that the CICR process cannot occur due to the complete absence of extracellular calcium, as well as the low probability of calcium release by the RyRs due to the blockade by procaine, eugenol was nevertheless able to induce significant contraction force.

The regulation of the excitation-contraction coupling process is orchestrated through a spatially defined program of ion channels and exchangers, which strictly control the entry of Ca^{2+} into the cell as well as the release of Ca^{2+} from the sarcoplasmic reticulum. RyRs are the Ca^{2+} channels most abundant inside the cardiac myocyte. The Ca^{2+} thus generated increases the contraction of the filaments through its interaction with troponin C and the subsequent increase in actin-myosin interaction (Williams et al., 1992; Minamikawa et al., 1997). Following this line of reasoning, we developed a protocol in which atrial trabeculae, after a period of stabilization and maintained under pacemaker stimulation, were treated with ryanodine (100 μM), a known RyR antagonist. Immediately after the addition of ryanodine, we observed a complete abolition of the isometric contraction force developed by the atrial muscle. When we added eugenol (5 mM) to this solution, it was able to induce a significant increase in resting tension. This eugenol-induced tension had the same amplitude as that developed by eugenol (5 mM) in the control condition. However, we repeated the same experimental procedures in zero-Ca ringer's solution^{2+} , to confirm the non-cooperation of extracellular Ca^{2+} in eugenol's ability to increase resting tension. Under these experimental conditions, we obtained the same results as those observed in normal ringer's solution. These results suggest that the action of eugenol is not via RyRs, but possibly via a Ca^{2+} independent pathway.

On the other hand, considering that the results reported so far are contradictory in relation to those described in the literature, we developed a new protocol in which the atrial trabeculae, after a period of stabilization and maintained under pacemaker stimulation in normal ringer, were treated with ryanodine (100 μM) and procaine (10 mM), both blockers of RyR activity. Again controlling the cooperation of extracellular Ca^{2+} , we carried out the experiments in normal ringer solution and zero-Ca ringer^{2+} . The results obtained in the protocols in which the blockers were used alone were repeated even in the presence of both, i.e. immediately after treatment, we observed a complete abolition of the force of the isometric contraction developed by the atrial trabeculae, but when eugenol (5 mM) was added to this preparation, this compound was able to induce an increase in resting tension.

BDM has a negative inotropic effect on both intact and permeabilized cardiac muscle (Li et al., 1985; Daly et al., 1987). This compound has no major effect either on the electrical activity of the heart or on the release of Ca^2 + by the SR (Mulieri & Alpert, 1984; Li et al., 1985). BDM reduces the sensitivity of myofilaments to Ca^{2+} , shifting the Ca^{2+} sensitivity curve to the right (Li et al., 1985). It also inhibits the formation of cross-bridges (Mulieri & Alpert, 1984; Blanchard et al., 1984). Therefore, myofibrils seem to be the main site of action of BDM. On the other hand, the action of BDM is concentration-dependent (Tripathy et al., 1999); since at low concentrations (<2 mM) it reduces force by reducing the Ca^2 + transient while at high concentrations (10 and 30 mM) it also affects the cyclization of cross-bridges, inhibiting them by keeping the AM.ADP.Pi state stabilized (Herrmann et al., 1992; Fryer et al., 1988), as well as decreasing the sensitivity of myofibrils to Ca^{2+} . In our experiments, we used 30 mM BDM in electrically stimulated trabeculae, 10 mM and 30 mM in trabeculae permeabilized by the action of saponin. Since one of the actions of BDM is to reduce transient Ca^{2+} , we carried out the protocol with intact fibers in normal ringer solution and zero-Ca ringer^{2+} , and in the presence of procaine. Our results on intact trabeculae showed that, after treatment with BDM, the increase in resting tension induced by eugenol was reversed. Similarly, in experiments with permeabilized atrial trabeculae, the results showed that when the trabeculae were treated with BDM (10 mM) plus eugenol (5 mM), we obtained a significant reduction in resting tension compared to that developed by the control condition (eugenol 5 mM). However, when the trabecula was treated with BDM (30 mM) plus eugenol (5 mM), we observed that eugenol was unable to induce an increase in resting tension. These results allow us to hypothesize that BDM in high concentrations (30 mM) was able to inhibit the cyclization of cross-bridges (Fryer, et al., 1988; Blanchard et al., 1990; Maylie & Hui, 1991; Herrmann, et al., 1992; Kagawa, et al., 1995) and, consequently, eugenol, under these experimental conditions, was unable to increase the resting tension.

Therefore, our results in preparations using BDM suggest that this compound, by inducing complete relaxation of the atrial trabecula, suggests that the action of eugenol is the result of an active state of cyclization of cross-bridges, and not a state of "*rigor mortis*".

The possibility of depleting the Ca^{2+} stored in the SR was observed by Bradley et al (2002) in smooth muscle cells. The SR was depleted of Ca^{2+} by the action of caffeine (10 mM) applied sequentially at intervals of 1 to 2 minutes. On the other hand, caffeine (30 mM) was used in experiments with crustacean muscles, resulting in the activation of RyRs and the consequent release of stored Ca^2 + (Lea, J. T., 1996). Caffeine has the property of binding effectively to RyRs, in other words, as long as there is caffeine in the solution these receptors remain activated, which enables the constant release of Ca^{2+} from the SR (Bradley et all, 2002). A second pathway has been proposed as a mechanism contributing to cardiac CICR, the *Na /*Ca^{+2+} *exchanger* (Bridge et al., 1990). This exchanger represents a mechanism by which Ca^{2+} movements through the sarcolemma are coupled to reciprocal $Na+$ movements, with a stoichiometry of $3Na+$:$1Ca^2$ + (Brenner, 1993). In our experiments, caffeine was able to promote the depletion of the SR, observed by the lack of response to the presence of caffeine in the solution. Even after this depletion, the trabecula, when exposed to Eugenol (5 mM), was able to increase its resting tension, a tension that was not the result of cooperation with the Na /Ca^{+2+} exchangers, confirming previous data that the action of eugenol is independent of RyRs and via a pathway independent of Ca^2 + ions.

The contractile capacity of the myocardium can be altered in one of three ways: 1) by varying the amount of Ca^2 + available to activate the myofilaments, 2) by altering the binding affinity of the filament to Ca^{2+} and, 3) by altering the kinetics of the filaments when binding Ca^{2+} (Ford, 1991). Experiments carried out using the membrane permeabilization technique (*"skinned fibers"*) allow the mechanical properties of contractile proteins to be investigated without interfering with the mechanisms of the AEC process. Studies on permeabilized fibers provide information on the active and passive mechanical properties of the myofibrillar apparatus (Matsubara et al., 1989).

The experiments with permeabilized fibers, therefore, in the absence of electrical stimulation, allow the properties of the cross-bridge cycle to be analyzed, since the production of the force observed is determined by the higher concentration of Ca^2 + ions. On the other hand, this preparation also makes it possible to determine the ability of the SR to release Ca^2 + ions in the presence of caffeine, a known RyR agonist (Vites and Pappano, 1994).

The sensitivity of the contractile apparatus to Ca^2 + ions can be determined when we use

Solution A (pCa 4.0). The results in this solution indicate the maximum force production capacity as a result of the maximum concentration of Ca^{2} + ions in the solution (Lofrano-Alves, M.S., et al., 2005).

When the atrial trabeculae were transferred to Solution E plus caffeine (30 mM), we could observe an increase in force production, which occurred as a result of the release of Ca^{2+} by the RyRs. These results corroborate those found in experiments with crustacean skeletal muscle (Lea, J. T., 1996). Likewise, these data confirm the viability of the SR membrane even after saponification.

Immediately afterwards, the preparation was transferred to Solution L in order to recharge the SR with Ca^{2} + ions. This time, the preparation was transferred to Solution E plus eugenol (5 mM). The results indicate that eugenol was able to induce a significantly higher force development when compared to that induced by caffeine (RyRs agonist).

However, studies on isolated skeletal muscle and cardiac muscle cells from which the sarcolemma has been mechanically removed (skinned fibers) make it possible to investigate the activation processes of the contractile system (Gordon, 1978). Removing the sarcolemma allows solutions containing ions and substrates to directly bathe the contractile filaments. In this way, the effects of ions and substrates on the muscle tension recordings of dismembered fibers can be used to evaluate their interactions with contractile proteins (Gordon, 1978). The solubilization of the lipid bilayer using different non-ionic detergents has attracted attention, particularly through the use of Triton X-100 (Almgren, 2000). The use of Triton X-100 has been used to permeabilize both the cell membrane (sarcolemma) and the membrane of organelles. This procedure allows us to study contractile proteins more directly (Almgren, 2000).

After obtaining the maximum tension due to the action of eugenol, Triton X-100 (1% vol/vol) was added to the bath solution. Under these experimental conditions, we observed that eugenol was unable to maintain the increase in resting tension, resulting in relaxation with a lower resting tension than the initial condition. These results suggest that eugenol is not acting directly on the contractile proteins, as already observed in the BDM experiments.

Considering, as seen above, the possibility that eugenol is not acting via RyRs, we could hypothesize that the force developed was induced by a pathway independent of RyRs and Ca ions^{2+} . These data confirm those found in experiments with intact muscle. However, when we interfere with the AEC mechanism, we are interfering with muscle contraction via calcium ions, as the specialized literature shows. Therefore, eugenol could be acting through or associated with compartmentalized calcium in a series of sites located inside and outside the cell, such as: extracellular calcium; calcium bound to the anionic sites of the glycocalyx; slow current calcium (DHPRs); calcium bound to the inner face of the membrane; calcium bound to the SR; myoplasmic calcium which acts on troponin; mitochondrial calcium and nuclear calcium. Ruthenium red is a water-soluble compound obtained by reacting RuCl3 and NH3 in solution (Joly, 1892). It is a complex compound with a high molecular weight and is therefore impermeable to membranes (Charuk et al., 1990). However, ruthenium red is a compound widely used in saponin-permeabilized fiber preparations, as it acts as a RyR-blocking agent (Kargacin et al., 1998; Marx et al., 2001). Ruthenium red has an inhibitory effect on the transport and release of Ca^{2+} from mitochondrial stores (Gunter, T. E., and Pfeiffer, D. R., 1990), is an inhibitor of the Ca^{2+} releasing channels of the SR in smooth muscle and skeletal muscle (Garcha, R. S., and Hughes A. D., 1994; Kanmura, Y, et al., 1989; Chamberlain, B. K., et al., 1984; Ma, J., 1993; Mack, M.M., et al., 1992); the plasma membrane Ca^{2+} -ATPase pump (PMCA) (Missiaen et al., 1990) the SR membrane Ca^{2+} pump (SERCA) (Kanmura et al., 1989); InsP3-induced Ca^{2+} release (therefore inhibits IP3Rs) (Kanmura et al., 1989; Vites & Pappano, 1994), as well as several Ca-binding proteins^{2+} such as calmodulin (Sasaki et al., 1992) and calsequestrin (Charuk et al., 1990). The results of eugenol in the presence of ruthenium red suggest that in saponin-permeabilized trabeculae, this essential oil was able to induce an increase in resting tension statistically equal to that developed by eugenol in the control condition (eugenol 5 mM). Thus, our results suggest that eugenol is acting through a pathway independent of the one described in the CICR process. These data corroborate those found both in experiments with intact atrial trabeculae and those found in experiments with saponin-permeabilized atrial trabeculae. These data were confirmed by the results obtained in the protocol in which we related the concentration of Ca^2 + in the solution to force production in the absence and presence of eugenol (5 mM). Under these experimental conditions it was possible to observe that there was no change in the force curve when comparing the results in the presence and absence of eugenol, indicating that this compound does not act by sensitizing contractile proteins to calcium ions.

In the permeabilized fiber technique, we can use solutions with different concentrations of EGTA, a well-known Ca chelator^2 + (Godt & Maughan, 1977). Under these conditions, we observed a complete absence of contractile response in the presence of caffeine (30 mM), which confirms that the 5 mM concentration of EGTA is sufficient to chelate all the Ca^{2+} released by the SR via RyRs. Next, after passing through Solution R (relaxant), the preparation was transferred to another chamber containing Solution R (high EGTA) plus eugenol (5 mM). This procedure showed us that eugenol was able to induce an increase in resting tension even in solution with EGTA (5 mM). These results demonstrate that eugenol is inducing an increase in resting tension in atrial trabeculae through a Ca-independent pathway^{2+} . Based on these results, we then proceeded to permeabilize this same preparation with Triton X-100. Thus, we obtained a preparation in which it would be possible to determine whether or not the compound under study (eugenol) was acting directly on the contractile myofilaments. At the end of this stage, we could see that eugenol was not able to induce an increase in resting tension as observed in previous protocols. Also in this protocol, the preparation was transferred back to Solution A (pCa 4.0). Under these conditions, the atrial trabeculae were able to develop contraction force, suggesting that Triton X-100 was not able to alter the cyclization capacity of the contractile filaments.

The experiments in which we used the technique of permeabilization of the piasmatic membrane by saponin make it possible to use agonists and antagonists for the mobilization and/or release of Ca^{2+}, which are impermeable to the plasma membrane (Matsubara et al., 1989). They also make it possible to identify the completeness of the SR membrane and associated proteins, such as RyRs, since we observed an increase in resting tension in the presence of caffeine (Vites and Pappano, 1994). It also allows the use of an effective Ca^{2+} chelator, EGTA (Godt & Maughan, 1977), and even under these conditions eugenol is able to increase the resting tension. However, when we chemically remove the plasma membrane and cell organelle membranes, including the SR, using Triton X-100 (Almagren, 2000), a technique that allows us to act directly on contractile proteins (Gordon, 1978), we observe a reversal of eugenol's ability to increase resting tension. To answer this question, we carried out a protocol in which we permeabilized the atrial trabeculae with saponin for 20, 40, 60 and 80 minutes, and also used intact trabeculae. The results showed that, in order to obtain maximum activation of the contractile proteins in solution A, permeabilization of the plasma membrane is necessary; that caffeine is able to cross the plasma membrane and that the response to this compound is increased when the trabecula is permeabilized, and as the response was maintained over time, it indicates that permeabilization by saponin has no effect on the SR membrane; However, the most significant result was that the increase in resting tension induced by eugenol decreased considerably over the time of saponin permeabilization, reaching 80 minutes without any response. These data are similar to those observed in the presence of Triton X-100, and indicate that the pathway by which eugenol acts is soluble and localized in the cytosol.

1,4-Dithiothreitol (DTT) is a reducing agent, routinely used as a protective agent in all experiments with enzymes and proteins. Studies on skeletal muscle cells have shown that the redox action of DTT does not activate the release of Ca^{2+} from the SR, and can also effectively inhibit the depolarization that induces the release of Ca^{2+} (Posterino & Lamb, 1996). Studies have shown that DTT does not alter the free Ca^{2+} concentration in intact muscle cells during submaximal tetany experiments, and that even under these experimental conditions DTT results in a decrease in force production (Andrade et al., 1998). In our experiments, after the maximum tension was obtained in solution A, and the caffeine was replaced in solution E, the trabeculae were incubated in solution R plus DTT (100 μM) for 20 minutes. After the incubation period, the trabeculae were transferred to solution R plus DTT (100 μM) and Eugenol (5 mM). The results show that after treatment with DTT and in the presence of DTT, eugenol was able to significantly increase the resting tension when compared to the resting tension developed by control eugenol. These results cannot be attributed to DTT's ability to develop strength, since Andrade et al (1998) have already shown that this is not possible. Thus, this significantly greater response is possibly the result of this component's protective capacity against protein degradation, reaffirming that the pathway by which eugenol acts is dependent on a cytoplasmic protein agent.

Colchicine is a very toxic alkaloid, extracted mainly from the bulbs of plants of the *Colchicum autumnale* species belonging to the *Liliaceae* family (Havas, 1937). It is also the substance most commonly used to induce polyploidy in genetic improvement programs for agricultural crops, forestry species and ornamental plants (Sharma & Sharma, 1999; Silva et al., 2000).

The mechanism of action of colchicine is known. It binds reversibly to the tubulin dimer, causing a conformational change that prevents polymerization of the mitotic spindle and consequently blocks the cell in metaphase. Since the specific sites to which colchicine binds in the α and β dimers are inaccessible when tubulin is polymerized in the form of microtubules, it is prevented from polymerizing. In already formed microtubules, colchicine prevents the growth of these structures during cell division (Morejohn et al., 1987; Sluder, 1991).

At lower concentrations, colchicine promotes the depolymerization of the mitotic spindle, due to the disorganization of the microtubules (Dolezel, 1995). When it binds with tubulin, it forms the *colchicine-tubulin* complex, which prevents the addition of tubulin at the plus (+) end of the microtubule, resulting in the disassembly of the microtubule (Junqueira & Carneiro, 2004).

Microtubules are structures found in the cytoplasm and also in cell extensions, such as cilia and flagella. They are rigid structures and play a significant role in the development and maintenance of cells (Junqueira & Carneiro, 2004), but they are also involved in the positioning of membrane-bound vesicles and organelles, and the transportation of membrane-bound vesicles is one of their main functions (Cooper, 2001). Tubulin polymerization during microtubule formation (polymerization and depolymerization) depends on the concentration of Ca2+ in the cytosol and the participation of Microtubule Associated Proteins or MAPs (Junqueira & Carneiro, 2004).

In our experiments, we induced depolymerization or disorganization of the microtubule by adding colchicine (30 μM) to the bath solution, leaving it there for 20 minutes, which was enough time for this compound to act. After incubation, we added eugenol (5 mM) to the solution, and under these experimental conditions, eugenol was able to induce an increase in resting tension greater than that observed with control eugenol. These results suggest, together with those observed in the presence of DTT, that the pathway by which eugenol acts is dependent on a soluble cytosolic protein agent associated with microtubules.

Apparently, the protective action of DTT and the potentiation observed by colchicine indicate that eugenol acts via a soluble cytosolic pathway. To test this hypothesis, we carried out a protocol to analyze the possibility of reincorporating this or these soluble agents into an atrial trabecula treated with Triton X-100. The lack of response in solution E plus caffeine (30 mM) confirmed that the SR membrane had been dissolved by the action of Triton X-100. As in previous protocols, under these experimental conditions, eugenol (5 mM) was unable to increase resting tension. When the preparation was transferred to the experimental solution called SUCO, it was able to induce an increase in resting tension corresponding to 20% of the tension obtained with control eugenol. When the preparation was transferred to solution R (SUCO) plus eugenol (5 mM), this was able to induce an increase in resting tension statistically equal to the tension in control eugenol. The same was observed in solution R (SUCO + COLCH), even without the presence of eugenol, the tension developed was equal to that observed in control eugenol. However, when we added eugenol to solution R (SUCO+COLCH), we obtained greater development of resting tension when compared to the other results. Taken together, these data indicate that the permeabilization product alone is capable of increasing the resting tension, which is substantially higher in the presence of eugenol, when permeabilization is carried out in the presence of colchicine, there is a potentiation of the increase in resting tension in the presence and absence of eugenol, confirming the previous data that eugenol acts via a cytosolic, soluble pathway, through proteins that may be associated with microtubules.

Sodium arsenite, or trivalent arsenic, is a widely used tool for studying the effects of p38 MAPk activation in cardiac myocytes. The predominant action of arsenite is a strong and sustained activation of p38 MAPk (Chen et al., 2003). Experiments carried out to identify the activity of p38 MAPk in intact hearts, using a transgenic approach, showed that acute activation of this kinase directly resulted in a depression of the capacity of the myofilaments to develop maximum force (Vahebi et al., 2007), 2007). The same was observed in permeabilized preparations, i.e. a reduction in the maximum force-producing capacity of myofilaments when p38 MAPk was stimulated by arsenite (Chen et al. 2003; Vahebi et al., 2007).

Studies were carried out with Isoeugenol, eugenol and alibenzene to investigate their possible effects on the signaling of a wide variety of kinases, including p38 MAPk. The results show that these essential oils, including eugenol, are able to inhibit the production of NO (nitric oxide) and the expression of iNOS, and that their effects are mediated by blocking the phosphorylation (inhibition) of ERK 1/2 and p38 MAPk, as well as the degradation of I-κBα, and activation of NF-kB (Choi et al., 2007).

Our results show that, in a dose-dependent manner, arsenite (1, 3, 5, 7 and 10 mM) was able to reverse the increase in resting tension induced by eugenol (5 mM). The dose-response curve enabled us to determine the concentration of arsenite to be used in the other protocols (IC50=6.78 mM). Even though arsenite is a toxic agent, experiments with intact and electrically stimulated atrial trabeculae showed that even after exposure to arsenite (7 mM) there was a recovery of contractile capacity (50% compared to twitch), and after the preparation was washed, eugenol was able to induce an increase in resting tension. In saponin-permeabilized preparations, after obtaining the maximum force in solution A, the trabecula was incubated with arsenite (7 mM) for 30 minutes, and under these experimental conditions, arsenite was not able to alter the sensitivity of contractile proteins to Ca2+, a result corroborated by the data presented by Vahebi et al (2007). However, eugenol's ability to induce an increase in resting tension was inhibited in the presence of arsenite (7 mM). These results were repeated when the preparation was treated with Triton X-100, i.e. there was no change in the sensitivity to Ca^{2+} by the contractile proteins, but again, arsenite was able to inhibit eugenol's ability to increase resting tension. Taken together, these data indicate that the pathway by which eugenol acts is possibly due to its ability to inhibit p38 MAPk activity, as already demonstrated by Choi et al (2007).

One of the properties of eugenol described in the literature refers to its ability to act directly or indirectly on the concentration of free ATP. In studies carried out with fibroblasts, it was observed that this oil was able, at concentrations of 1 and 2 mM, incubated for 1 hour, to reduce the cellular ATP concentration by 18% and 46%, respectively, suggesting that eugenol's ability to alter the free ATP concentration is time and concentration dependent (Jeng et al., 1994). Another study investigated the possible effects of eugenol on the F0, F1 subunits of mitochondrial membrane ATPase. The results showed that eugenol was able to stimulate the hydrolysis of ATP, as well as inhibiting the oxidation of NADH, resulting in a reduction in the free concentration of ATP (Usta et al., 2002). The activation of myosin filaments induced by thin filaments was studied in isolated cardiac myocytes, as well as in isolated skeletal muscle fibers. In these cases, it was observed that the number of cross-bridges formed gradually increased in response to the gradual reduction in the concentration of MgATP in solution, even in the absence of Ca^{2+} . The results of this study indicate that the formation of cross-bridges, at least in the absence of Ca2+, is the result of cooperative activation triggered by the thin filament, further concluding that in the absence of Ca2+, cooperative activation is greater in cardiac muscle than skeletal muscle (Metzger, 1995; Kad et al., 2005).

In a very similar way to that observed in the literature, in our experiments eugenol was able to significantly reduce the concentration of ATP in extracts of atrial trabeculae in a time-dependent manner. In this way, another of the possible ways in which eugenol acts may be related to the oil's ability to reduce the free concentration of ATP, resulting in an increase in resting tension through the cooperative activation of cardiac thin filaments, forming cross-bridges in a state of strong bonding.

CONCLUSIONS

The results presented here show that eugenol's ability to induce an increase in resting tension is independent of the pathways for releasing Ca^2 + from cellular stores and the mobilization of extracellular Ca +.^{2}

The increase in resting tension induced by eugenol is not a consequence of a state of *"rigor mortis"* since BDM was able to reverse this tension.

Eugenol does not act directly on contractile proteins by extracting troponin I or directly phosphorylating myosin, since after permeabilization with Triton X-100, this oil was unable to induce an increase in resting tension.

In permeabilized preparations, the increase in resting tension induced by eugenol is the result of its ability to inhibit p38 MAPk, while in intact preparations the induction of an increase in resting tension is the result of its ability to reduce the free concentration of ATP, thus forming cooperative cross-bridges.

Taken together, these results show that eugenol does not act specifically by a single route, but regardless of the route by which it acts, the result is always an increase in resting tension.

In this way, the use of eugenol as an experimental tool can help us to describe these pathways, which may be involved in a wide variety of alterations in cardiac function.

BIBLIOGRAPHICAL REFERENCES

Ahlijamian MK; Westenbroek RE; Catterall WA. Subunit structure and localization of dihydropyridine-sensitive calcium channels in mammalian brain, spinal cord, and retina. Neuron. 1990; 4: 819-32.

Aires, MM. Fisiologia Humana. 2ed. Rio de Janeiro: Guanabara Koogan, 1999. 795p.

Almeida, M.A. Effects of eugenol on guinea pig tracheal smooth muscle. Fortaleza, 2004. 127p. Dissertation (Master's Degree in Physiological Sciences) - Health Sciences Center, State University of Ceará.

Almers, W., Stirling, C. Distribution of transport proteins over animal cell membranes. J. Memb. Biol, 1984; 77: 169-186

Almgren, M. Mixed micelles and other structures in the solubilization of bilayer lipid membranes by surfactants, Bioch. et Bioph. Acta, 2000; 1508:146-163.

Andrews MAV et al, Ion specific and general ionic effects on contraction of skinned fast-twitch skeletal muscle from the rabbit. J. of Gen. Phy., 1991; 98: 1105-1125.

Bagni MA, Cecchi G, Colomo F, Garzella P. Effects of 2,3-butanedione monoxime on the crossbridge kinetics in frog single muscle fibres. J. Muscle. Res. Cell. Motil. 1992; 13: 516-22.

Bangalore, R., Mehrke, G., Gingrich, K., Hofmann, F., Kass, R.S. Influence of L- type Ca channel $\alpha 2/\delta$-subunit on ionic and gating current in transiently transfected HEK 293 cells. Am. J. Physiol, 1996; 270: H1521-H1528.

Bennett, PB. Morphological aspects of extracellular polysaccharides. J. Hist. and Cytoch. 1963; 11: 14.

Berlin, J.R. Spatiotemporal changes of Ca^{2+} during electrically evoked contractions in atrial and ventricular cells. Am J Physiol, 1995; 269:H1165- H1170.

Berridge, M.J.; Irvine, R.F.; Inositol trisphosphate, a novel second messenger in cellular signal transduction Nature. Sep 21;341(6239):197-205, 1989.

Bers, DM. Cardiac excitation-contraction coupling. Nature, 2002; 45:198-205.

Bers, DM. Excitation-contraction coupling and cardiac contractile force. 2st ed. Dordrecht, Netherlands, Kluwer Academic Publishers, 2001.

Bers, DM; Stiffel, VM. Ratio of ryanodine to dihydropyridine receptors in cardiac and skeletal muscle and implications for E-C coupling. American Journal of Physiology, 1993; 264:

C1587-C1593.

Bezprozvanny, I., Watras, J., Ehrlich, B.E.. Bell-shaped calcium-response curves of Ins(1,4,5)P3- and calcium-gated channels from endoplasmic reticulum of cerebellum. Nature, 1991; 351:751,754.

Bezprozvanny, I; Ehrlich, BE. The inositol 1, 4, 5-trsiphosphate (InsP) receptor. J Membr Biol, 1995; 145:205-216.

Blanchard, EM; Smith, GL; Allen, DG; Alpert, NR; The effects of 2,3-butanedione monoxime on initial heat, tension, and aequorin light output of ferret papillaey muscles. Pflug Arch Eur J, 1990; 416:219-221.

Blanchard, EM; Solaro, RJ; Inhibition of the activation and troponin calcium binding of dog cardiac myofibrils by acidic pH. Circ Res, 1984; 55:382-391.

Block, BA; Imagawa, T; Campbell, KP; Franzini-Armstrong, C. Structural evidence for direct interaction between the molecular components of the transverse tubule/sarcoplasmic reticulum junction in skeletal muscle. Journal of Cellular Biology, 1988; 107: 2587-2600.

Blondel, O; Takeda, J; Jansen, H; Seino, S and Bell, GI. Sequence and functional characterization of a third inositol trisphosphate receptor subtype, IP3R-3, expressed in pancreatic islet, kidney, gastrointestinal tract and other tissues. J Biol Chem, 1993; 268:11356-11363.

Bodi, I; Mikala, G; Koch, SE; Akhter, AS; Schwartz, A. The L-type calcium channel in the heart: the beat goes on. The Journal of Clinical Investigation, 2005; 115.

Boehning D; Mak DOD; Foskett JK; Joseph SK; Molecular determinants of ion permeation and selectivity in inositol 1,4,5-trisphosphate receptor Ca^{2+} channels. J Biol Chem, 200; 1276: 13509-13512.

Bootman, M.D.; Lipp, P & Berridge, M.J. The organization and functions of local Ca^{2+} signals. J Cell Sci, 2001; 114:2213-2222.

Bootman, MD; Collins, TJ; Mackenzie, L; Roderick, HL; Berridge, MJ; Peppiatt, CM;. 2-Aminoethoxydiphenyl borate (2-APB) is a reliable blocker of store- operated Ca^{2+} entry, but an inconsistent inhibitor of InsP3-induced Ca^{2+} release, FASEB J., 2002; 16:1145-1150.

Bradley, K.N; Flynn, E.R.M; Muir, T.C; McCarron, J.G. Ca2+ regulation in guineapig colonic smooth muscle: the role of the Na+-Ca2+ exchanger and the sarcoplasmic reticulum. *Journal of Physiology* (2002), 538.2, *pp.* 465-482

Bremel, RD & Weber, A. Calcium binding to rabbit skeletal myosin under physiological conditions. Biochemistry Biophysics Acta, 1975; 376: 366-374.

Brenner, B. Effect of Ca^{2+} cross-bridge turnover kinetics in skinned single rabbit psoas fibers. Proceedings of the National Academy of Sciences, USA, 1988; 85: 3265-33269.

Brenner, B; Yu, LC; Structural changes in the actomyosin cross-bridges associated with force generation. Proc Natl Acad Sci (USA), 1993; 90:52525256.

Bridge, J.H.B., Smolley, J.R., Spitzer, K.W.. Isolation of the sodium-calcium exchange current underlying sodium-dependent relaxation in heart muscle. Science, 1990; 248: 376-378.

Brilla C; Janicki JS; Weber KT. Impaired diastolic function and coronary reserve in genetic hypertension. Role of interstitial fibrosis and medial thickening of intramyocardial coronary arteries. Circulation Research, 1991; 69: 107-115.

Brodin, P & Roed, A. Effects of eugenol on rat phrenic nerve and phrenic nervediaphragm preparations. Archives of Oral Biology, 1984; 29(8): 611-615.

Brown, J.H., Jones, L.G.. Phosphoinositide metabolism in the heart, in Phosphoinositides and receptor mechanisms. Putney, J.W. Jr (ed). New York, NY, 1986; 245-270.

Burgess, G.M., Bird, G.S., Obie, J.F., Putney, J.W Jr. The mechanism for synergism between phospholipa. J. Biol. Chem., 1991; 266: 4772-4781.

Cameron, AM; Steiner, JP; Sabatini, DM; Kaplin, AL; Walensky, LD; Snyder, SH. Immunophilin FK506 protein associated with inositol 1,4,5-triphosphate receptor modulates calcium flux. Proc. Natl. Acad. Sci. USA, 1995; 92:17841788.

Carl, S.L.; Felix, K.; Caswell, A.H.; Brandt, N.R.; Ball, W.J.Jr; Vaghy,P.L.; Meissner, G. & Ferguson, D.G. Immunolocalization of sarcolemmal dihydropyridine receptor and sarcoplasmic reticulum triadin and ryanodine receptor in rabbit ventricle and atrium. J Cell Biol, 1995; 129:672-682.

Caroni P & Carafoli, E. An ATP-dependent Ca^{2+} -pumping system in dog heart sarcolemma. Nature, 1980; 283: 765-767.

Chadwick, CC; Saito, A and Fleischer, S. Isolation and characterization of the inositol trisphosphate receptor from smooth muscle. Proc Natl Acad Sci USA, 1990; 87:2132-2136.

Chamberlain BK, Volpe P, Fleischer S. Inhibition of calcium-induced calcium release from purified cardiac sarcoplasmic reticulum vesicles. J Biol Chem, 1984; 259:7547-7553

Chamberlain, BK; Volpe, P; Fleischer S. Inhibition of calcium-induced calcium release from purified cardiac sarcoplasmic reticulum vesicles. J Biol Chem, 1984; 259:7547-7553.

Chapman, RA. Control of cardiac contractility at the cellular level. American Journal of Physiology. Heart Circ Physiol,1983; 14: H535-H552.

Charuk, JHM; Pirraglia, CA; Reithemeier, RAF. Interaction of ruthenium red with Ca^{2+} -binding proteins. Analytical Biochemis, 1990; 188:123-131.

Chen, Y; Ravi, R; Qinghang, L; and Polly, H; Acute p38 MAPK activation decreases force development in ventricular myocytes. Am. Journal Physiology Heart, 285: H2578-H2586, 2003

Cheng HP, Lederer WJ. Calcium sparks. Physiol. Rev. 2008; 88: 1491-545.

Cheng HP, Wang SQ. Calcium signaling between sarcolemmal calcium channels and ryanodine receptors in heart cells. Front. Biosci. 2002; 7: D1867-D1878.

Cheng, H; Lederer, WJ; Cannell, MB. Calcium sparks: elementary events undelying excitatory-contraction coupling in heart muscle. Science, 1993; 262: 740-744.

Choi, C.Y.; Park, Kyung-Ran; Lee, Jung-Hee; Jeon, Y.J.; Liu, K.H.; Oh, S.; Kim, D.E.; Yea, S.S.; Isoeugenol suppression of inducible nitric oxide synthase expression is mediated by down-regulation of NF-κB, ERK ½, and p38 Kinase. European Journal of Pharmacology, 576; 151-159; 2007.

Cingolani, HE; Houssay, AB. Houssay's Human Physiology. 7ª edition. Sao Paulo: Artmed publishing house, 2004.

Clapham, DE. Calcium signaling. Cell, 1995; 80: 259-268.

Clifford MN. Miscellaneous phenols in foods and beverages - nature, occurrence and dietary burden. J. Sci. Food. Agric. 2000; 80: 1126-37.

Constantin, LL & Podolsky, RJ. Apud Fabiato, A & Fabiato, F. Calcium release from the sarcoplasmic reticulum. Circ Res, 1977; 40: 119-129.

Cooper, G. M. The cell: a molecular approach. Editora Artimed, 2ª edition: 2001.

Corbalan-Garcia, S.; Teruel, A; Gomez-Fernandez, J.C. Characterization of Ruthenium Red-binding sites of the Ca^{2+} -ATPase from sarcoplasmic reticulum and their interaction with Ca^{2+} -Binding sites. Murcia: J. Biochem, 1992; 287, 767-774

Craveiro, AA; Rodrigues, AS; Andrade, CHS. Volatile constituents of brazilian Euphorbiaceae genus Croton. J Nat Prod, 1981; 44(5): 509-632.

Curry, FE. Determinants of capillary permeability: a review of mechanisms based on single capillary studies in the frog. Circ Res, 1986; 59: 367-380.

Daly, MJ; Elz, JS; Nayler, WG. Contracture and the calcium paradox in the rat heart. Circ Res, 1987; 61:560-569.

Damiani CEN, Moreira CM, Zhang HT, Creazzo TL, Vassallo DV. Effects of eugenol, an essential oil, on the mechanical and electrical activities of cardiac muscle. J. Cardiovasc. Pharmacol. 2004; 44: 688-95.

Damiani, CEN; Moreira, CM; Zhang, HT; Creazzo, TL and Vassallo, DV. Effects of eugenol, an essential oil, on the mechanical and electrical activities of cardiac muscle. J Card Pharmacol, 2004; 44:688-695.

De Jongh, KS; Merrick, DK; Catterall, WA. Subunits of purified calcium channels. A 212-kDa form of α_1 and partial amino acid sequence of a phosphorylation site of an independent β subunit. Proce of the Nat Acad of Sci, USA, 1989; 86: 8585-8589.

De Waard, LJ; Pragnell, M; Campbell, KP. Ca^{2+} channel regulation by a conserved β subunit domain. Neuron, 1994; 13:495-503.

Del Nido, PJ; Glynn, P; Buenaventura, P; Salama, G; Koretsky, AP. Fluorescence measurement of calcium transients in perfused rabbit heart using rhod 2. The Am Physiol Society, 1998; 43:728-741.

Del Valle-Rodriguez, A; Lopez-Barneo, J; Urena, J. Ca^{2+} channel-sarcoplasmic reticulum coupling: a mechanism of arterial myocyte contraction without Ca^{2+} influx. EMBO J., 2003; 22:1-9.

Dip, E.C; Pereira, N.A; Fernandes, P.D.C. Ability of eugenol to reduce tongue edema induced by *Dieffenbachia picta* Schott in mice. Toxic. v. 43, p. 729735, 2004.

Dolezel, J. Dolezelovâ, M.; Novak, F. J. Flow cytometric estimation of nuclear DNA amount in diploid bananas (Musa acuminata and M. balbisiana). Biologia Plantarum, Dordrecth, v. 36, p. 351-357, 1994.

Ducke, A. Estudos botânicos no Cearâ. Annals of the Brazilian Academy of Sciences, 1959; 31: 231.

Dulhunty AF. Excitation-contraction coupling from the 1950s into the new millennium. Clin and Exp Pharmacol and Physiol, 2006; 33, 763-772.

Elison,C.; Jenden,D.J. Effects of Ryanodine on Model Systems Derived from Muscle. Bioch Pharmacol, 1967; 16:1339.

Endo, M. Calcium release from the sarcoplasmic reticulum. Physiol Rev, 1977; 57: 71-108.

Endo, M. Mechanism of action of caffeine on the sarcoplasmic reticulum of skeletal muscle. Proc. Jpn. Acad, 1975; 51, 479-484.

Endoh, M. Cardiac α1-adrenoceptors that regulate contractile function: Subtypes and subcellular signal transduction mechanisms. Neuroch Res, 1996; 21: 217-229.

Ertl, R; Jahnel, U; Nawrath, H; Carmeliet, E; Vereecke, J; Differential electrophysiologic and inotropic effects of phenylephrine in atrial and ventricular heart muscle preparations from rats. Naunyn Schmiedebergs Arch Pharmacol, 1991; 344:574-581.

Escobar, R. G. Eugenol: Pharmacological and toxicological properties. Advantages and disadvantages of its use. Ver Cubana Estomatol, v. 39, 2002.

Fabiato, A & Fabiato, F. Calcium-induced release of calcium from the cardiac sarcoplasmic reticulum. Am J of Physiol, 1983; 245: c14.

Fabiato, A. & Fabiato, F. Contractions induced by a calcium-triggered release of calcium from sarcoplasmic reticulum of single skinned cardiac cells. J Physiol, London, 1975; 249: 469-495.

Fabiato, A. Comparison and relation between inositol (1, 4, 5)-trisphosphate- induced release and calcium-induced release of calcium from the sarcoplasmic reticulum, in Recent Advances in Calcium Channels and Calcium Antagonists. Yamada K; Shibata S (eds). Elmsford, NY, Pergamon Press, Inc. 1990; 35-39.

Fabiato, A. Dependence of the Ca^{2+} -induced release from the sarcoplasmic reticulum of skinned skeletal muscle fibres forming the frog semitendinosus on the rate of change of free Ca^{2+} concentration at the outer surface of the sarcoplasmic reticulum. J Physiol, 1984; 353-56.

Fabiato, A. Inositol (1, 4, 5)-trisphosphate-induced release of Ca^{2+} from the sarcoplasmic reticulum of skinned cardiac cells. Biophys J, 1986a; 49:190a.

Fabiato, A. Inositol (1, 4, 5)-trisphosphate-induced versus Ca^{2+} from the cardiac sarcoplasmic reticulum. Proc Int Union Physiol Sci, 1986b; 16:350.

Fabiato, A. Rapid ionic modifications during the aequorin-detected calcium transient in a skinned canine cardiac Purkinje cell. J Gen Physiol, 1985; 85:189-246.

Fabiato, A & Fabiato, F. Calcium release from the sarcoplasmic reticulum Circ Res, 1977; 40: 119-129.

Fairhurst.AS. Effect of Ryanodine on Skeletal-Muscle Reticulum Calcium Adenosine-Triphosphatase. Biochem Pharmacol, 1973; 22:2815-2827.

Farah, CS; Reinach, FC. The troponin complex and regulation of muscle contraction. FASEB. J., 1995; 9:755-767.

Ferrier, G.R.; Howlett, S.E. Cardiac excitation-contraction coupling: role of membrane potential in regulation of contraction. American Journal of Physiology-Heart and Circ Physiol, 2001; 280:H1928-H1944.

Flicker, PF; Phillips GN Jr; Cohen, C. Troponin and its interactions with tropomyosin. An electron microscope study. J Mol Biol, 1982; 162: 495-501.

Ford, L.E. Mechanical Manifestations of Activation in Cardiac-Muscle. Circ Res, 1991; 68:621-637.

Francisco H. Andrade, F.H; Reid, M.B; Allen, D.G; Westerblad, H.A. Effect of hydrogen peroxide and dithiothreitol on contractile function of single skeletal muscle fibres from the mouse. Journal of Physiology (1998), 509.2, pp. 565- 575 565

Franzini-Armstrong, C. Membrane particles and transmission at the triad. Fed Proc, 1975; 34: 1382-1389.

Franzini-Armstrong, C. Studies of the triad. Structure of the junction in frog twitch fibers. J Cell Biol, 1970; 47: 488-499.

Franzini-Armstrong, C; Protasi, F. Ryanodine receptors of striated muscles: A complex channel capable of multiple interactions. Physiol Rev, 1997; 77: 699-729.

Freiburg, A; Gautel, M. A molecular map of the interactions between titin and myosin-binding protein C. Implications for sarcomeric assembly in familial hypertrophic cardiomyopathy. Eur. J. Biochem, 1996; 235:317-323.

Fryer, MW; Neering, IR; Stephenson, DG. Effects of 2,3-butanedione monoxime on the contractile activation properties of fast- and slow-twitch rat muscle fibers. J. Physiol, 1988; 407:53-75.

Fuchs, F. The binding of calcium to glycerinated muscle fibers in rigor. The effect of filament overlap. Biochim. Biophys. Acta., 1977; 491:523-531.

Furuichi, T; Yoshikawa, S; Miyawaki, A; Wada, K; Maeda, N and Mikoshiba, K. Primary structure and functional expression of the inositol 1, 4, 5- trisphosphate-binding protein P400. Nature, 1989; 342:32-38.

Gafni J, Munsch JA, Lam TH, Catlin MC, Costa LG, Molinski TF & Pessah IN. Xestospongins: potent memebrane permeable blockers of the inositol 1,4,5- trisphosphate receptor. Neuron, 1997; 19:723-733.

Galvan, DL; Borrego-Diaz, E; Perez, PJ; Mignery, GA;. Subunit oligomerization, and topology of the inositol 1,4,5-triphosphate receptor. J. Biol. Chem, 1999; 274:29483-29492.

Gambassi, G., Spurgeon, H. A., Ziman, B. D., Lakatta, E. G. & Capogrossi, M. C. Opposing effects of α1-adrenergic receptor subtypes on Ca^{2+} and pH homeostasis in rat cardiac myocytes. Am. J. Physiol, 1998; 274, H1152- H1162.

Garcha RS; Hughes AD. Action of heparin and ruthenium red on responses of reversibly-permeabilized rat mesenteric arteries. Eur J Pharmacol, 1994; 268:319-325.

Gautel, M; Goulding, D; A molecular map of titin/connectin elasticity reveals two different mechanisms acting in series. FEBS Lett, 1996; 385:11-14.

Go, LO; Moschella, MC; Watras, J; Handa, KK; Fyfe, BS and Marks, AR. Differential regulation of two types of intracellular calcium release channels during end-stage heart failure. J Clin Invest, 1995; 95:888-894.

Godt RE & Lindley BD. Influence of temperature upon contractile activation and isometric force production in mechanically skinned muscle fibers of the frog. J Gen Physiol, 1982; 80: 279-297.

Godt RE; Fogaça RTH et al, Changes in force and calcium sensitivity in the developing avian heart. Can J Physiol and Pharmacol, 1991; 69: 1692-1697.

Godt, RE; Maughan, DW. Influence of osmotic compression on calcium activation and tension in skinned muscle fibers of the rabbit. Pflug. Arch., 1981; 391:334-337.

Godt, RE; Maughan, DW. Swelling of skinned muscle fibers of the frog, experimental observations. Biophys. J., 1977; 19:103-116.

Gordon, AM; Huxley, AF; Julian, FJ. The variation in isometric tension with sarcomere length in vertebrate muscle fibres. J Physiol, 1966; 184: 170-192.

Gordon, AR. Contraction of Detergent-treated smooth muscle. Proc. Natl. Acad. Sci. (USA), 1978; 75:3527-3530.

Gregory, RB; Rychkov, G; Barritt, GJ; Evidence that 2-aminoethoxydiphenyl borate is a novel inhibitor of store-operated Ca^{2+} channels in liver cells, and acts through a mechanism which does not involve inositol trisphosphate receptors, Biochem. J., 2001; 354:285-290.

Grynkiewicz, G; Poenie, M; Tsien, RY. A new generation of Ca^{2+} indicators with greatly improved fluorescence properties. J. Biol. Chem., 1985; 260:34403450.

Gunter TE, Pfeiffer DR. Mechanisms by which mitochondria transport calcium. Am J Physiol, 1990; 258:C755-C786

Gyork, S; Fill, M. Ryanodine receptor adaptation: Control mechanism of Ca^{2+} - induced Ca^{2+} release in heart. Science, 1993; 260:807-809.

Hackett, PH; Rodriguez, G; Roach, RC. Clove cigarettes and high-altitude pulmonary edema. [JAMA, 1985; 253: 3551-3552.

Hagan, EC; Jenner, PM; Jones, WI. Toxic properties of compounds related to safrole. Toxicol and Appl Pharmacol, 1965; 7: 18-24.

Hagar, RE; Ehrkucg, BE; Regulation of the type III InsP3 receptor by InsP3 and ATP. Biophys J., 2000; 79:271-278.

Han H-M, Robinson RB, Bilezikian JP & Steinberg SF. Developmental changes in guanine nucleotide regulatory proteins in the rat myocardial α1-adrenergic receptor complex. Circulation Res, 1989; 65:1763-1773.

Hancock, WO; Martyn, DA; Huntsman, LL. Ca^{2+} and segment length dependence of isometric force kinetics in intact ferret cardiac muscle. Circulation Res, 1993; 603-611.

Hartmann, HA; Mazzocca, NJ; Kleiman RB; Houser, SR. Effects of phenylephrine on calcium current and contractility of feline ventricular myocytes. Am J Physiol, 1988; 255: H1173-H1180.

Havas, L. Effects of colchicine and "vascum album" preparationa on germination of seeds and growth of seedlings. Nature, London, v. 139, n. 3513, p. 371372, 1937.

Herrmann, C.; Wray, J.; Travers, F. & Barman, T. Effect of 2,3-butanedione monoxime on myosin and myofibrillar ATPase. An example of an uncompetitive inhibitor. Biochem, 1992; 31, 12227-12232,.

Hescheler, J; Nawrath, H; Tang, M; Tratwein, W. Adrenoceptor-mediated changes of excitation and contraction in ventricular heart muscle from guineapigs and rabbits. J Physiol, 1988; 397: 657-670.

Higuchi, H. & Takemori, S. Butanedione monoxime suppresses contraction and ATPase activity of rabbit skeletal muscle. The J Biochem, 1989; 105, 638643.

Hirata, M; Suematsu, E; Hashimoto, T; Hamachi, T and Koga, T; Release of Ca^{2+} from non-mitochondrial store in peritoneal macrophages treated with saponin by inositol 1, 4, 5-trisphosphate. Biochem J, 1984; 223:229-236.

Hirs, A. Sales of clove cigarettes in US: health questions emerge. Tabacco International, 1984; 28: 57-58.

Holroyde, MJ; Robertson, SP; Johnson, JD et al, The calcium and magnesium binding sites on cardiac troponin and their role in the regulation of myofibrillar adenosine triphosphate. J Biol Chem, 1980; 255: 11668-11693.

Horowitz, A; Menice, CB; Laporte, R and Morgan, K. Mechanism of smooth muscle contraction. Physiol Rev, 1996; 76:967-1003.

Hui, C. S. Pharmacological studies of charge movement in frog skeletal muscle. J Physiol, 1983; 337:509-529.

Hume, W. R. Effect of eugenol on constrictor responses in blood vessels of the rabbit ear. J. Dent. Res., v. 62(9), p. 1013-5, 1983.

Hüser J; Lipsius SL & Blatter LA. Calcium gradients during excitation-contraction coupling in cat atrial myocytes. J Physiol, 1996; 494:641-651.

INTERNATIONAL CARNIVOROUS PLANT SOCIETY (ICPS). Available at: <http://www.carnivorousplants.org/howto/Propagation/Colchicine.php>.

Accessed on: March 10, 2012.

Inui, M; Saito, A; Fleischer S. Purification of the ryanodine receptor and identity with feet tructures of junctional terminal cisternae of sarcoplasmic reticulum from fast skeletal muscle. J Biol Chemi, 1987a; 262: 1740-1747.

Inui, M; Saito, A; Fleischer, S. Isolation of the ryanodine receptor from cardiac sarcoplasmic reticulum and identity with the feet structures. J Biol Chem, 1987b; 262: 15637-15642.

Ivanov, S.V., et al. Cerebellar ataxia, seizures, premature death, and cardiac abnormalities in mice with targeted disruption of the Cacna2d2 gene. Am. J. Pathol, 2004; 165:1007-1018.

Jahnel U, Duwe E, Pfennigsdorf S & Nawrath H. On the mechanism of action of phenylephrine in rat atrial heart muscle. Naunyn Schmeidebergs Arch Pharmacol, 1994; 349:408-415.

Jahnel U, Jakob H & Nawrath H. Electrophysiologic and inotropic effects of alpha-adrenoceptor stimulation in human isolated atrial heart muscle. Naunyn Schmeidebergs Arch Pharmacol, 1992a; 346:82-87.

Jahnel U, Nawrath H, Carmeliet E & Vereecke J. Depolarization-induced influx of sodium in response to phenylephrine in rat atrial heart muscle. J Physiol. 1991; 432:621-637.

Jahnel U, Nawrath H, Shieh RC, Sharma VK,Williford DJ & Sheu SS. Modulation of cytosolic free calcium concentration by alpha 1-adrenoceptors in rat atrial cells. Naunyn Schmeidebergs Arch Pharmacol. 1992b; 346:88-93.

Jenden, DJ; Fairhurst, AS; The pharmacology of ryanodine. Pharmacol Rev, 1969; 21:1-25.

Jeng,LJ.Hahn,FJ.Lu,Y.J.Wang,and M.Y.P.Kuo. Eugenol Triggers Different Pathobiological Effects on Human Oral Mucosal Fibroblasts[1] . J Dent Res. 1994; 73(5): 1050-1055.

Joly, A. Compose's ammoniacaux de'rive's du Sesquichlorure de ruthe'nium. C R Acad Sci, 1892; 115:1299-1301.

Jonas, S., Sugimori, M., and Llinas, R. Is low molecular weight heparin a neuroprotectant? Ann. N.Y. Acad. Sci.1997; 825:389-393

Junqueira, L.C; Carneiro, J. Basic Histology. Editora Guanabara Koogan, 10ª edition: 2004.

Kad NM, Kim S, Warshaw DM, VanBuren P, Baker JE. Single-myosin crossbridge interactions with actin filaments regulated by troponintropomyosin. Proc. Natl. Acad. Sci. U S A 2005; 102: 16990-5.

Kagawa, K; Horiuti, K; Yamada, K. BDM compared with P(i) and low Ca^{2+} in the cross-bridge reaction initiated by flash photolysis of caged ATP. Biophys. J, 1995; 69:2590-2600.

Kanmura Y; Raeymaekers L; Casteels R. Effects of doxorubicin and ruthenium red on intracellular Ca^{2+} stores in skinned rabbit mesenteric smooth-muscle fibers. Cell Calcium, 1989; 10:433-439.

Kargacin, G.J.; Ali, Z; Kargacin, M.E. (1998). Ruthenium red reduces the Ca^{2+} sensitivity of Ca^{2+} uptake into cardiac sarcoplasmic reticulum. Pflügers Arch - Eur J Physiol (1998) 436:338-342

Katz, AM. A growth of ideas: Role of calcium as activator of cardiac contraction. Card Res, 2001; 52:8-13

Katz, AM. Molecular basis of calcium channel blockade. Am J Cardiol, 1992; 69: 17E-22E.

Katz, AM. Purification and properties of tropomyosin-containing protein fraction that sensitizes reconstituted actomyosin to calcium-binding agents. J Biol Chem, 1966b; 241: 1522-1529.

Katz, AM. Role of the contractile proteins and sarcoplasmic reticulum in the response of the heart to catecholamine: a historical review. Adv Cyclic Nucleot Res, 1979; 11: 303-343.

Katz, AM; Repke, DI; Cohen, B. Control of the activity of highly purified cardiac actomyosin by Ca^{2+} , Na^{+} , and K^{+} . Circ Res, 1966a; 19: 1062-1070.

Katz, AM; Takenaka, H and Watras, J. The sarcoplasmic reticulum, in The heart and Cardiovascular System. Fozzard HA (ed). New York, NY, Raven Press, 1986; 731-746.

Kentish, JC; Barsotti, RJ; Lea, TJ; Mulligan, IP; Patel, JR and Ferenczi, MA. Calcium release from sarcoplasmic reticulum induced by photorelease of calcium or Ins(1, 4, 5)$_{P3}$. Am J Physiol, 1990; 258:H610-H615.

Keung EC & Karliner JS. Complex regulation of calcium current in cardiac cells; dependence on a pertussis toxinsensitive substrate, adenosine triphosphate, and an α1-adrenoceptor. J Clin Invest, 1990; 85:950-954.

Kirsch, CM; Yenokida, GG; Jensen, WA; Wendland, R; Suh, H; Bourgault, M. Non-cardiogenic pulmonary edema due to the intravenous administration of clove oil. Thorax, 1990; 45: 235-236.

Kobayashi, S., Somlyo, A.V., Somlyo, A.P. Heparin inhibits the inositol 1,4,5- trisphosphate-dependent, but not the independent, calcium release induced by guanine nucleotide in vascular smooth muscle. Biochem. Biophys. Res. Commun, 1988; 153, 625-631.

Kockskamper,J.; Sheehan,K.A.; Bare,D.J.; Lipsius,S.L.; Mignery,G.A.; Blatter,L.A. Activation and propagation of Ca^{2+} release during excitationcontraction coupling in atrial myocytes. Biophy J, 2001; 81:2590-2605.

Komai H, Rusy BF. Effects of inhibition of trans-sarcolemmal calcium influx by nickel on force of postrest contraction and on contracture induced by rapid cool. Cardiovasc Res, 1993; 27:801-806.

Kozam, G & Mantell, GM. The effect of eugenol on oral mucous membrane. Journal of Dental Research, 1978; 57: 954-957.

Lahlou, S; Caldas Magalhaes, P.J; Leal-Cardoso, J.H; Duarte, G.P. Cardiovascular Effects of Eugenol, a Phenolic Compound Present in Many Plant Essential Oils, in Normotensive Rats. J. of Cardiov. Pharmac. v. 43(2) p. 250-257, 2004.

Lai, FA; Erickson, HF; Rousseau, E; Liu, QY; Meeissner, G. Purification and reconstitution of the calcium release channel from frog skeletal muscle. Nature, 1988; 331: 315-319.

Lamb, GD; Posterino, GS and Stephenson, DG. Effects of heparin on excitatory contraction coupling in skeletal-muscle of toad and rat. J Physiol, 1994; 474:319-329.

Launikonis,B.S.; Stephenson,D.G. Effect of saponin treatment on the sarcoplasmic reticulum of rat, cane toad and crustacean (yabby) skeletal muscle. Journal of Physiology-London, 1997; 504:425-437.

Lavabre, M. Aromatherapia: Healing through essential oils. 4ª ed. Rio de Janeiro: Record, 1997.

Lavoie,E.J.; Adams,J.D.; Reinhardt,J.; Rivenson,A.; Hoffmann,D. Toxicity Studies on Clove Cigarette-Smoke and Constituents of Clove - Determination of the Ld50 of Eugenol by Intratracheal Instillation in Rats and Hamsters. Archives of Toxicology, 1986; 59:78-81.

Lea,T.J. Caffeine and micromolar Ca^{2+} concentrations can release Ca^{2+} from ryanodine-sensitive stores in crab and lobster striated muscle fibres. Journal of Experimental Biology, 1996; 199:2419-2428.

Leal-Cardoso, JH; Coelho-de-Souza, AN; Souza, IT; Figueiredo, IM. Effects of eugenol on excitation-contraction coupling in skeletal muscle. Archives Internationales de Pharmacodynamie et de Thérapie, 1994; 327(1): 113-124.

Lehrer, SS; Golitsina, NL; Geeves, MA. Actin-tropomyosin activation of myosin subfragment-1 ATPase and thin filament cooperativity. The role of tropomyosin flexibility and end-to-end interactions. Biochemistry, 1997; 36: 13449-13454.

Li K, He H, Li C, Sirois P & Rouleau JL. Myocardial α1-adrenoceptor: inotropic effect and physiologic and pathologic implications. Life Sc, 1997; 60:13051318.

Li, L; DeSantiago, J; Chu, G; Kranias, EG; Bers, DM. Phosphorylation of phospholamban and troponin I in β-adrenergic-induced acceleration of cardiac relaxation. American Journal of Physiology, 2000; 278: H769-H779.

Li, T.; Sperelakis, N.; Teineck, R. E.; & Solaro, J. R. Effects of diacetyl monoxime on cardiac excitation-contraction coupling. J. Pharmacol. Exp. Ther, 1984; 232, 688-695.

Lipp, P; Hüser, J; Pott, L. & Niggli, E. Spatially non-uniform Ca^{2+} signals induced by reduction of transverse tubules in citrate-loaded guinea-pig ventricular myocytes in culture. J. Physiol, 1996a; 497:589-597.

Lipp, P; Hüser, J; Pott, L. & Niggli, E. Subcellular properties of triggered Ca^{2+} waves in isolated citrate-loaded guinea-pig atrial myocytes characterized by ratiometric confocal microscopy. J Physiol, 1996b; 497:599-610.

Lipp, P; Laine, M; Tovey, SC; Burrell, KM; Berridge, MJ; Li, W and Bootman, MD. Functional InsP3 receptors that may modulate excitation-contraction coupling in the heart. Current

Biology, 2000; 10:939-942.

Lipp, P; Pott, L; Callewaert, G & Carmeliet, E. Simultaneous recording of Indo-1 fluorescence and Na /Ca^{+2+} exchange current reveals two components of Ca^{2+} release from sarcoplasmic reticulum of cardiac atrial myocytes. FEBS Lett, 1990; 275:181-184.

Lofrano-Alves, MS; Oliveira, EL; Damiani, CEN; Kassouf-Silva, I and Fogaça, RTH; Eugenol-induced contractions of saponin-skinned fibers are inhibited by heparin or by a ryanodine receptor blocker. Canadian Journal of Physiology and Pharmacology, 2005; 83:1-8.

Lorell, BH; Apstein, CS; Cunningham, MJ; Schoen, FJ; Weiberg, EO; Peeters, GA; Barry, WH. Contribution of endothelial cells to calcium-dependent fluorescence transients in rabbit hearts loaded with indo-1. Circ. Res, 1990; 67:415-425.

Ma J. Block by ruthenium red of the ryanodine-activated calcium release channel of skeletal muscle. J Gen Physiol, 1993; 102:1031-1056.

Mack MM; Zimânyi I; Pessah IN. Discrimination of multiple binding sites for antagonists of the calcium release channel complex of skeletal and cardiac sarcoplasmic reticulum. J Pharmacol Exp Ther, 1992; 262:1028-1037.

Mackenzie L; Bootman MD; Laine M; Berridge MJ; Thuring J; Holmes A; Li WH; Lipp P. The role of inositol 1,4,5-trisphosphate receptors in Ca^{2+} signalling and the generation of arrhythmias in rat atrial myocytes. J Physiol, 2002; 541: 395-409.

Mackenzie, L.; Roderick, H.L.; Berridge, M.J.; Conway, S.J. & Bootman, M.D. The spatial pattern of atrial cardiomyocyte calcium signaling modulates contraction. J Cell Sci, 2004; 117:6327-6337.

MacMillan, D; Chalmers, S; Muir, TC; McCarron, JG. IP3-mediated Ca^{2+} increases do not involve the ryanodine receptor, but ryanodine receptor antagonists reduce IP3-mediated Ca^{2+} increases in guinea-pig colonic smooth muscle cells. J. Physiol; 2005; 569.2:533-544.

Magyar J, Szentandrassy N, Banyasz T, Fulop L, Varro A, Nanasi PP. Effects of terpenoid phenol derivatives on calcium current in canine and human ventricular cardiomyocytes. Eur. J. Pharmacol. 2004; 487: 29-36.

Maranto AR. Primary structure, ligand binding, localization of the human type 3 inositol 1,4,5-trisphosphate receptor expressed in intestinal epithelium. J Biol Chem, 1994; 269: 1222-1230.

Marchant, J.S. & Parker, I. Functional interactions in Ca^{2+} signaling over different time and distance scales. J Gen Physiol, 2000; 116:691-696.

Maruyama, T; Kamaji, T; Nakada, S; Kanno, T; Mikoshiba, K; 2APB, 2-aminoethoxydiphenyl borate, a membrane-penetrable modulator of Ins(1,4,5)P3-induced Ca^{2+} release. J Biochem, 1997; 5005:498-505.

Marx, SO; Gaburjakova, J; Gaburjakova, M; Henrikson, C; Ondrias, K; Marks, AR. Coupled gating between cardiac calcium release channels (ryanodine receptors). Circ Res, 2001; 88:1151-1158.

Matsubara,I.; Maughan,D.W.; Saeki,Y.; Yagi,N. Cross-Bridge Movement in Rat Cardiac-Muscle As A Function of Calcium-Concentration. Journal of Physiology-London, 1989; 417:555-565.

Maylie, J; Hui, CS; Action of 2,3-butanedione monoxime on calcium signals in frog cut twitch fibers containing Anipyrylazo-III. J Physiol, 1991; 442:551-567.

Meissner, G. Isolation and characterization of two types of sarcoplasmic reticulum vesicles. Biochemistry, Biophysics Acta, 1975; 389: 51-68.

Meissner, G. Ryanodine activation and inhibition of the Ca^{2+} release channel of sarcoplasmic reticulum. Journal of Biological Chemistry, 1986; 261 (14): 6300-6306.

Metzger JM. Myosin binding-induced cooperative activation of the thin filament in cardiac myocytes and skeletal-muscle fibers. Biophys. J. 1995; 68: 1430-42.

Michailesco P, Kouassi M, El Briak H, Armynot A, Boudeville P. Antimicrobial activity and tightness of a DCPD-CaO-based hydraulic calcium phosphate cement for root canal filling. J. of Biom. Mat. Res. Part B-Applied Biomaterials 2005; 74B: 760-7.

Michikawa, T; Hamanaka, H; Otsu, H; Yamamoto, A; Miyawaki, A; Furuichi, T; Tashiro, Y; Mikoshiba, K; Transmembrane topology and sites of N- glycosylation of inositol 1,4,5-triphosphate receptor. J. Biol. Chem, 1994; 269:9184-9189.

Mignery, GA; Südhof, TC; Takei, K and Camilli, P. Putative receptor for inositol 1, 4, 5-trisphosphate similar to ryanodine receptor. Nature, 1989; 342:192195.

Mignery, GA; Sudhof, TC; The ligand binding site and transduction mechanism in the inositol-1,4,5,-triphosphate receptor. EMBO J., 1990; 9:3893-3898.

Minamikawa, T., Cody, S., & Williams, D. In situ visualization of spontaneous calcium waves within perfused whole rat heart by confocal imaging. Am J Physiol, 1997; 272:H236-H243.

Missiaen, L.; Callewaert, G.; De Smedt, H.; and Parys, J. B. 2- aminoethoxydiphenyl borate affects the inositol 1,4,5-trisphosphate receptor, the intracellular Ca^{2+} pump and the non-specific leak from the non- mitochondrial Ca^{2+} stores in permeabilized A7r5 cells. Cell Calcium, 2001; 29:111-116

Missiaen, L; Callewaert, G; De Smedt, H; Parys, JB;. 2-Aminoethoxydiphenyl borate affects the inositol, 1,4,5-trisphosphate receptor, the intracellular Ca^{2+} pump and the non-specific leak from the non-mitochondrial Ca^{2+} stores in permeabilized A7r5 cells, Cell Calcium, 2001 ; 29:111-116.

Moore, ED; Voigt, T; Kobayashi, YM; Isenberg, G; Fay, FS; Gallitelli, MF; Franzini-Armstrong, C. Organization of Ca^{2+} release units in excitable smooth muscle of the guinea-pig urinary bladder. Biophysical Journal, 2004; 87:1836-1847.

Moraru II; Kaftan EJ; Ehrlich BE; Watras J; Regulation of type 1 inositol 1,4,5- trisphosphate-gated calcium channels by InsP3 and calcium: simulation of single channel kinetics based on ligand binding and electrophysiological analysis. J Gen Physiol, 1999; 113: 837-849.

Morejohn, L. C.; Bureau, T. E.; Bajer, M.; Bager, A. S.; Fosket, D. E. Oryzalin, a dinitroaniline herbicide, binds to plant tubulin and inhibits microtubule polymerization in vitro. Planta, Berlin, v. 172, n. 2, p. 252-264, 1987.

Morris SA; Nerou EP; Riley AM; Potter BV; Taylor CW; Determinants of adenophostin A binding to inositol trisphosphate receptors. Biochem, 2002; J 367: 113-120.

Moschella,M.C.; Marks,A.R.. Inositol 1,4,5-Trisphosphate Receptor Expression in Cardiac Myocytes. Journal of Cell Biology, 1993; 120:1137-1146

Mulieri, LA; Alpert, NR. Differential effects of 2,3-butanedione monoxime (BDM) on activation and cantraction. Biophys., 1984; 45-47.

Nishijima, H; Uchida, R; K; Kawakami, N; Shimamura, K; Kitamura, K. Role of endothelium and adventia on eugenol-induced relaxation of rabbit ear artery precontracted by histamine. Journal of Smooth Muscle Research, 1998; 34: 123-137.

Nishijima, H; Uchida, R; Kameyama, K; Kawakami, N; Ohkubo, T; Kitamura, K. Mechanisms mediating the vasorelaxing ation of eugenol, a pungent oil, on

rabbit arterial tissue. Jpn. J. Pharmacol, v. 79, p. 327-334. Nature 312, 315 - 321, 1999.

Nomura, N; Satoh, H; Terada, H; Hayashi, H. CAM II is responsible for reactivation of SR Ca uptake and contractile recovery during intracellular acidosis. Journal of Molecular Cell Cardiology, 2000; 32: A98.

Nosek, TM; Willians, MF; Ziegler, ST and Godt, RE. Inositol trisphosphates enhances calcium release in skinned cardiac and skeletal muscle. Am J Physiol, 1986; 250:C807-C811.

O'Neill, SC; Miller, L; Hinch, R; Eisner, DA. Interplay between SERCA and sarcolemmal Ca^{2+} efflux pathways controls spontaneous release of Ca^{2+} from the sarcoplasmic reticulum in rat ventricular myocytes. J. Physiol., 2004; 559.1:121-12.

Obermann, WM; Plessmann, U; Weber, K; Furst, DO. Purification and biochemical characterization of myomesin, a myosin-binding and titin-binding protein from bovine skeletal muscle. European Journal of Bichemistry, 1995; 233: 110-115.

Offer, G. C-protein and periodicity in the thick filaments of vertebrate skeletal muscle. Cold Spring Harbor Symposium Quant Biol, 1972; 37: 87-93.

Ogawa, Y. Role of ryanodine receptors.Crit Rew Biochem Mol Biol. 1994; 29:229274.

Ohkubo, T & Kitamura, K. Eugenol activates Ca^{2+} -permeable currents in rat dorsal root ganglion cells. Journal of Dental Research, 1997; 76(11): 17371744.

Ohtsuki, I. Molecular arrangement of troponin T in the thin filament. Journal of Biochemistry, 1979; 86: 491-497.

Olivoto, R.R. Effects of eugenol on the excitation-contraction coupling (ECC) process of rat cardiac muscle: involvement of the IP3 receptor (IP3R) pathway. Curitiba, 2008. 150 p. Dissertation (Master's Degree in Cellular and Molecular Biology - Physiology) - Center for Biological Sciences, Federal University of Paranà.

Opie, LH. Regulation of myocardial contractility. Journal of Carciovascular Pharmacology, 1995; 26: S1-S9.

Ostwald, TJ & MacLennan, DH. Isolation of high affinity calcium-binding protein from sarcoplasmic reticulum. Journal of Biological Chemistry, 1974; 249: 974-979.

Ozeki, M. The effects of eugenol on the nerve and muscle in crayfish. Comparative Biochemistry and Physiology, 1975; 50C: 183-191.

Page, E. Quantitative ultrastructural analysis in cardiac membrane physiology. American Journal of Physiology, 1978; 235: C147-C158.

Page, E; McCallister, LP; Power, B. Stereological measurements of cardiac ultrastructures implicated in excitation-contraction-coupling. Proceedings of the National Academy of Sciences, USA, 1971; 68: 1465-1466.

Peppiatt, MC; Collins, TJ; Mackenzie, L; Conway, SJ; Holmes, AB;, Bootman, MD; Berridge,

MJ; Seo, JT; Roderick, HL;. 2-Aminoethoxydiphenyl borate (2- APB) antagonizes inositol 1,4,5-trisphosphate-induced calcium release, inhibits calcium pumps and has a use-dependent and slowly reversible action on store-operated calcium entry channels. Cell Calcium, 2003; 97-108.

Perez DM, DeYoung MB & Graham RM. Coupling of expressed α1B- and α1D- adrenergic receptors to multiple signaling pathways is both G protein and cell type specific. Mol Pharmacol, 1993; 44:784-795.

Perez, PJ; Ramos-Franco, J; Fill, M and Mignery, G. Identification and functional reconstitution of the type 2 Inositol 1, 4, 5-Trisphosphate receptor from ventricular cardiac myocytes. The Journal of Biological Chemistry, 1997; 19:23961-23969.

Perez-Reyes, E & Schneider, T. Molecular biology of calcium channels. Kidney Int, 1995; 48: 1111-1124.

Pessoa, L.M; Morais, S.M; Bevilaqua, C.M.L; Luciano, J.H.S. Anthelmintic activity of essential oil of Ocimum gratissimum Linn. And Eugenol against Haemonchus contortus. Veterin. Parasit. v. 109, p. 59-63, 2002.

Piacentino, V. III, Dipla, K., Gaughan, J. P. & Houser, S. R. Voltage-dependent Ca^{2+} release from the SR of feline ventricular myocytes is explained by Ca^{2+} - induced Ca^{2} + release. J. Physiol, 2000; 523, 533-548

Poggioli, J., Sulpice, J. C. & Vassort, G. Inositol phosphate production following a1-adrenergic, muscarinic, or electrical stimulation in isolated rat heart. FEBS Lett, 1986; 206, 292-298.

Pogwizd, SM; Schlotthauer, K; Li, L; Yuan, W & Bers, DM. Arrhythmogenesis and contractile dysfunction in heart failure: roles of sodium-calcium exchange, inward rectifier potassium current and residual b-adrenergic responsiveness. Circ. Re, 2001; 88, 1159-1167.

Posterino, G. S. & Lamb, G. D. Effects of reducing agents and oxidants on excitation-contraction coupling in skeletal muscle fibres of rat and toad. Journal of Physiology 496, 809-825. 1996.

Potocnik, SJ; Hill, MA; Pharmacological evidence for capacitative Ca^{2+} entry in cannulated and pressurized skeletal muscle arterioles, Br. J. Pharmacol, 2001; 134:247-256.

Ramos-Franco, J; Fill, M; Mignery, GA; Isoform-specific function of single inositol 1,4,5-triphosphate receptor channels. Biophys J., 1998; 75:834-839.

Ramos-Franco, J; Galvan, D; Mignery, GA; Fill, M; Location of the permeation pathway in the recombinant type I inositol 1,4,5-triphosphate receptor. J Gen Physiol, 1999; 114:234-250.

Rayment, I; Holden, HM; Whittaker, M; Yohn, CB; Lorenz, M; Holmes, KC; Milligan, RA. Structure of actin-myosin complex and its implications for muscle contraction. Science, 1993 a; 261: 58-65.

Rayment, I; Holden, HM; Whittaker, M; Yohn, CB; Lorenz, M; Holmes, KC; Milligan, RA. Structure of actin-myosin complex and its implications for muscle contraction. Science, 1993 b; 261: 58-65.

Ringer, S. A further contribution regarding the influence of the different constituents of the blood on the contraction of the heart. Journal of Physiology, 1883; 4: 29-47.

Roberts JF, Attari N, Sherriff M. The survival of resin modified glass ionomer and stainless steel crown restorations in primary molars, placed in a specialist paediatric dental practice. Br. Dent. J. 2005; 198: 427-31.

Rüegg, JC. Towards a molecular understanding of contractility. Cardioscience, 1990; 1: 163-167.

Sasaki, T; Inui, M; Kimura, Y; Kuzuya, T; Tada, M. Molecular mechanism of regulation of Ca2_ pump ATPase by phospholamban in cardiac sarcoplasmic reticulum. Effects of synthetic phospholamban peptides on Ca^{2+} -pump ATPase. J Biol Chem, 1992; 267:1674-1679.

Scales, DJ. Aspects of the mammalian cardiac sarcotubular system revealed by freeze-fracture electron microscopy. Journal of Molecular Cell Cardiology, 1981; 13: 373-380.

Scholz, J; Schaeffer, B; Schmitz, W; Scholz, H; Steinfath, M; Lohse, M; Schwabe, U and Puurunen, J. α1-adrenoreceptor-mediated positive inotropic effect and inositol trisphosphate increase in mammalian heart. J Pharmacol Exp Ther, 1988; 245:337-345.

Schouten,V.J.A.; Morad,M. Regulation of Ca^{2+} Current in Frog Ventricular Myocytes by the Holding Potential, Camp and Frequency. Pflugers Archiv- European Journal of Physiology, 1989; 415:1-11.

Schumann HJ,Wagner J, Knorr A, Reidemeister JC, Sadony V & Schramm G. Demonstration in human atrial preparations of α-adrenoceptors mediating positive inotropic effects. Naunyn Schmiedebergs Arch Pharmocol, 1978; 302:333-336.

Sensch O, Vierling W, Brandt W, Reiter M. Effects of inhibition of calcium and potassium currents in guinea-pig cardiac contraction: comparison of betacaryophyllene oxide, eugenol, and nifedipine. Br. J. Pharmacol. 2000; 131: 1089-96.

Sensch, O; Vierling, W; Brandt, W; Reiter, M. Effects of inhibition of calcium and potassium currents in guinea-pig cardiac contraction: comparison of β- caryophyllene oxide, eugenol, and nifedipine. British Journal of Pharmacology, 131, 1089 ±1096, 2000.

Sham, JSK; Hatem, SN; Morad, M. Species differences in the activity of the Na^{+} - Ca^{2+} exchanger in nmammalian cardiac myocytes. Journal of Physiology, 1995; 488: 623-631.

Sharma, A. K.; Sharma, A. Plant chromosomes: analysis, manipulation and engineering. Amsterdam: Hardwood Academic, 1999. 371 p.

Sheehan,K.A.; Blatter,L.A. Regulation of junctional and non-junctional sarcoplasmic reticulum calcium release in excitation-contraction coupling in cat atrial myocytes. Journal of Physiology-London, 2003; 546:119-135.

Silva, S. de O. e; Rocha, S.A.; Alves, E. J.; Credico, M.; Passos, A. R. Morphological characterization and evaluation of banana cultivars and hybrids. Revista Brasileira de Fruticultura, Jaboticabal, v. 22, p. 161-169, 2000.

Singer, D; Biel, M; Lotan, I; Flockerzi, V; Hofamann, F; Dascal, N; The roles of the subunits in the function of the calcium channel. Science, 1991; 253:15531557.

Singer,S.J.; Nicolson,G.L. Fluid Mosaic Model of Structure of Cell-Membranes. Science, 1972: 175:720.

Skomedal T, Aass H, Osnes JB, Fjeld NB, Klingen G, Langslet A & Semb G. Demonstration of alpha adrenoceptormediated inotropic effect of norepinephrine in human atria. J Pharmacol Exp Therapeutics, 1985; 233:441-446.

Sluder, G. The practical use of colchicine and colcemid to reversibly block microtubule assembly in living cells. In: K. Adolph, editor. Marcel Dekker Inc., Advanced Techniques in Chromosome Research. New York. p. 427-447, 1991.

Smith, JS; Rousseau, E and Meissner, G. Calmodulin modulation of single sarcoplasmic reticulum Ca release channels from cardiac and skeletal muscle. Circ Res, 1989; 64:352-359.

Solaro, RJ & Rarick, HM. Troponin and tropomyosin-proteins that switch on and tune in the activity of cardiac myofilaments. Circ. Res, 1998; 83, 471-480.

Solovyova, N. Pernyhough, P; Glazner, G; Verkhratsky, A. Xestospongin C empties the ER calcium stores but does not Inhibit InsP3-induced Ca^{2+} release in cultured dorsal root ganglia neurons. Cell Calcium, 2002; 32:4952.

Sperelakis, N; Katsube, Y; Yokoshiki, H; Sada, H; Sumii, K. Regulation of the slow Ca^{2+} channels of myocardial cells. Molecular and Cellular Biochemistry, 1996; 163-164: 85-98.

Steinberg SF, Drugge ED, Bilezikian JP & Robinson RB. Acquisition by innervated cardiac myocytes of a pertussis toxin-specific regulatory protein linked to the alpha 1-receptor. Science, 1985; 230:186-188.

Stephenson, D. G & West, J. M. Contractile activation and the effects of 2,3- butanedione (BDM) in skinned preparations from normal and dystrophic mice. Pflug. Arch-Eur. J. Physiol. 413, 546-552, 1989.

Stephenson, DG; Wendt, IR. Effects of procaine on calcium accumulation by the sarcoplasmic reticulum of mechanically disrupted rat cardiac muscle. J. Physiol, 1986; 373:195-207.

Stewart, PS; & MacLennan, DH. Surface particles of sarcoplasmic reticulum membranes. Strucutral features of the adenosine triphosphatase. Journal of Biological Chemistry; 1974, 249: 985-993.

Sticht, FD & Smith, RM. Eugenol: some pharmacologic observations. Journal of Dental Research, 1971; 50: 1531-1535.

Südhof, T; Newton, C; Archer, B; Ushkaryov, Y and Mignery, G. Structure of a novel InsP3 receptor. EMBO (Eur Mol Biol Orgn), 1991; 11:3199-3206.

Supattapone, S; Worley, PF; Baraban, JM and Snyder, SH. Solubilization, purification and characterization of an inositol trisphosphate receptor. J Biol Chem, 1988; 263:1530-1534.

Taylor CW; da Fonseca PC; Morris EP. IP3 receptors: the search for structure. Trends Biochem Sci, 2004; 29: 210-219.

Terzic A, Puceat M, Clement O, Scamps F & Vassort G. Alpha 1-adrenergic effects on intracellular pH and calcium and on myofilaments in single rat cardiac cells. J Physiol, 1992; 447:275-292.

Thompson D, Norbeck K, Olsson LI, Constantinteodosiu D, Vanderzee J, Moldeus P. Peroxidase-catalyzed oxidation of eugenol - formation of a cytotoxic metabolite(s). J. Biol. Chem. 1989; 264: 1016-21.

Thompson, D & Eling, T. Mechanism of inhibition of prostaglandin H synthase by eugenol and other phenolic peroxidase substrates. Molecular Pharmacology, 1989; 36: 809-817.

Thompson, D; Norbeck, K; Olsson, L-I; Constantin-Teodosiu, D; van der Zee, J; Moldéus, P. Peroxidase-catalyzed oxidation of eugenol: Formation of a cytotoxic metabolite(s). The

Journal of Biological Chemistry, 1989; 264(2): 1016-1021.

Tripathy,A.; Xu,L.; Pasek,D.A.; Meissner,G. Effects of 2,3-butanedione 2- monoxime on Ca^{2+} release channels (ryanodine receptors) of cardiac and skeletal muscle. Journal of Membrane Biology, 1999; 169:189-198.

Tsien, RW & Tsien, RY. Calcium channels, stores, and oscillations. Annual Reviews in Cell Biology, 1990; 6:7715-760.

Usta J, Kreydiyyeh S, Bajakian K, Nakkash-Chmaisse H. In vitro effect of eugenol and cinnamaldehyde on membrane potential and respiratory chain complexes in isolated rat liver mitochondria. Food. Chem. Toxicol. 2002; 40: 935-40.

Vahebi, S; Ota, A; Manxiang, L; Chad, M.W; Pieter P. T; Yibin, W; Solaro, R.J; p38-MAPK Induced Dephosphorylation of _-Tropomyosin Is Associated With Depression of Myocardial Sarcomeric Tension and ATPase Activity. *Circ Res.* 2007;100:408-415

Vaughan-Jones, RD. Excitation and contraction in heart: the role of calcium. Bristh Medical Bulletin, 1986; 42: 413-420.

Vergara, J. and C. Caputo. Effects of tetracaine on charge movements and calcium signals in frog skeletal muscle fibers. Proceedings of the National Academy of Sciences, USA, 1983; 80:14~77-1,481.

Vites A-M, Pappano AJ; Distinct modes of inhibition by ruthenium red and ryanodine of calcium-induced calcium release in avian atrium. J Pharmacol Exp Ther, 1994; 268:1476-1484

Wang,Y.G.; Dedkova,E.N.; Ji,X.; Blatter,L.A.; Lipsius,S.L. Phenylephrine acts via IP3-dependent intracellular NO release to stimulate L-type Ca^{2+} current in cat atrial myocytes. Journal of Physiology-London, 2005; 567:143-157.

Warshaw, DM. The in vitro motility assay: A window into the myosin molecular motor. News in Physiological Sciences, 1996; 11: 1-7.

Wei, XY; Pan, S; Lang, WH; Kim HY; Schneider, T; Perez-Reyes, E; Birnbaumer, L; Molecular determinants of cardiac Ca^{2+} channel pharmacology - Subunit requirement for the high affinity and allosteric regulation of of dihydropyridine binding. J Biol Chem, 1995; 270: 27106-27111.

Wendt-Gallitelli, M.F. Presystolic calcium-loading of the sarcoplasmic reticulum influences time to peak force of contraction. X-ray microanalysis on rapidly frozen guinea-pig ventricular muscle preparation. Basic Res. Cardiol. 80, 617±625, 1985.

Wibo, M; Bravo, G; Godfraind, T. Postnatal maturation of excitation-contraction coupling in rat ventricle in relation to the subcellular localization and surface density of 1,4-dihydropyridine and ryanodine receptors. Circulation Research, 1991; 68: 662-673.

Wier, W. G. & Balke, C. W. Ca^{2+} release mechanisms, Ca^{2+} sparks, and local control of excitationcontraction coupling in normal heart muscle. Circ. Res. 1999; 85, 770-776.

Williams AJ; West DJ; Sitsapesan R; Light at the end of the Ca^{2+} -release channel tunnel: structures and mechanisms involved in ion translocation in ryanodine receptor channels. Q Rev Biophys, 2001; 34: 61-104.

Williams, JH; Ward, CW; Reduced Ca^{2+} -induced Ca^{2+} release from skeletal muscle sarcoplasmic reticulum at low pH. Can J Physiol Pharmacol, 1992; 70:926-930.

Winegrad, S. Autoradiographic studies of intracellular calcium in frog skeletal muscle. Journal of General Physiology, 1965; 48: 455-479.

Woodcock, EA; Matkovich, SJ. Ins(1,4,5)P3 receptor and inositol phosphates in the heart - evolutionary artifacts or active signal transducers. Pharmacology & Therapeutics, 2005; 107:240-251.

Wrigh, SE; Baron, DA; Heffner, JE. Intravenous eugenol causes hemorrhagic lung edema in rats: Proposed oxidant mechanisms. Journal of Laboratory and Clinical Medicine, 1995; 125: 257-264.

Wu,X.; Zhang,T.; Bossuyt,J.; Li,X.D.; McKinsey,T.A.; Dedman,J.R.; Olson,E.N.; Chen,J.; Brown,J.H.; Bers,D.M. Local InsP(3)-dependent perinuclear Ca^{2+} signaling in cardiac myocyte excitation-transcription coupling. Journal of Clinical Investigation, 2006; 116:675-682.

Xu, L; Lai, FA; Cohn, A; Etter, E; Guerrero, A; Fay, FS; Meissner, G. Evidence for a Ca(21)-gated ryanodine-sensitive Ca^{2+} release channel in visceral smooth muscle. Proc. Natl. Acad. Sci., 1994; 91:3294-3298.

Yamada, M; Miyawaki, A; Saito, K; Nakajima, T; Yamamoto-Hino, M; Ryo, Y; Furuichi, T; Mikoshiba, K; The calmodulin-binding domain in the mouse type 1 inositol 1,4,5-triphosphate receptor. Biochem J, 1995; 308:83-88.

Yang,J.; McBride,S.; Mak,D.O.D.; Vardi,N.; Palczewski,K.; Haeseleer,F.; Foskett,J.K. Identification of a family of calcium sensors as protein ligands of inositol trisphosphate receptor Ca^{2+} release channels. Proceedings of the National Academy of Sciences of the United States of America, 2002; 99:7711-7716.

Yokoyama, C.T., et al. Mechanism of SNARE protein binding and regulation of Cav^2 channels by phosphorylation of the synaptic protein interaction site. Mol. Cell. Neurosci. 2005; 28:1-17.

Yoshikawa F; Morita M; Monkawa T; Michikawa T; Furuichi T; Mikoshiba K; Mutational analysis of the ligand binding site of the inositol 1,4,5- trisphosphate receptor. J Biol Chem, 1996; 271: 18277-18284.

Yue, DT; Marban, E; Wier, WG. Relationship between force and intracellular Ca^{2+} in tetanized mammalian muscle. Journal of General Physiology, 1986; 87: 223-242.

Zahradnikovà A, Palade P. Procaine effects on single sarcoplasmic reticulum Ca^{2+} release channel. Biophys J, 1993; 64:991-1003.

Zima, AV; Blatter, LA; Inositol-1,4,5-triphosphate-dependent Ca^{2+} signaling in cat atrial excitation-contraction coupling and arrhythmias. J Physiol, 2004; 555:607-615.

Zot, AS & Potter, JD. Structural aspects of troponin-tropomyosin regulation of skeletal muscle contraction. Annual Review Biophysic and Chemistry, 1987; 16: 535-559.

MIX
Papier aus verantwortungsvollen Quellen
Paper from responsible sources
FSC® C105338

Printed by Books on Demand GmbH, Norderstedt / Germany